American Institute of Chemical Engineers

Computational Molecular Science and Engineering Forum

Presentations at the
2007 AIChE Annual Meeting

November 4-9, 2007
Salt Lake City, Utah, USA

Printed from e-media with permission by:

Curran Associates, Inc.
57 Morehouse Lane
Red Hook, NY 12571
www.proceedings.com

ISBN: 978-1-60560-016-1

Some format issues inherent in the e-media version may also appear in this print version.

ISBN: 978-1-60560-016-1

American Institute of Chemical Engineers
Proceedings
Three Park Avenue
New York, NY 10016-5991
Phone: 212-591-8100

www.aiche.org

American Institute of Chemical Engineers

Computational Molecular Science and Engineering Forum
2007

TABLE OF CONTENTS

179a Self Assembly of Dipolar Colloidal Particles

Carol K. Hall, Chemical and Biomolecular Engineering, North Carolina State University, 911 Partners Way, Raleigh, NC 27695

To Be Announced

Web Page: turbo.che.ncsu.edu

179b Surface Curvature Effects On The Self-Assembly Of Surfactants On Colloidal Silica And In Silica Nanopores

G.H. Findenegg[1], T. Shin[1], and D. Lugo[2]. (1) Institut für Chemie, Stranski Lab, Technische Universität Berlin, Strasse des 17. Juni 124, Berlin 10623, Germany, (2) Institut für Chemie, Stranski Lab, Technische Universität Berlin, Strasse des 17. Juni 124, Berlin 10623, Germany

The adsorption of surfactants onto colloidal particles and in narrow pores plays an important role in many technological fields including surfactant-mediated ad-solubilization, or the removal of hydrophobic organics from aqueous media by adsorption and microfiltration processes. Adsorption isotherms of surfactants at hydrophilic surfaces commonly exhibit a sigmoidal shape with an initial low-affinity region, followed by a steep increase at a concentration somewhat below the CMC, indicating surface aggregation. Depending on the relative strength of the polar interactions of the surfactant heads with the surface, and the hydrophobic interactions among the surfactant tails, two limiting types of adsorbate structures can be formed: (a) laterally uniform films, or (b) surface micelles of various shapes. Structural information about adsorbed surfactant layers at flat solid/water interfaces can be obtained by neutron reflectometry and grazing incidence small-angle neutron scattering, and the structure of surfactant aggregates on colloidal silica can been studied by SANS. On the other hand, little is known about the structure of surfactant aggregates in nano-size pores, when confinement and curvature effects become important. We are studying the self-assembly of surfactants in the cylindrical pores of MCM-41 type periodic mesoporous silica by adsorption measurements and small-angle diffraction. The diffraction data can be analyzed in terms of structural models based on the structure factor of a 2D hexagonal lattice and the form factor of the cylindrical pore without or with an adsorbed film. Neutron scattering offers the possibility to choose grossly different contrast scenarios of the surfactant aggregates and water in the pores by using H2O/D2O mixtures of different scattering length densities. In favorable cases the Bragg reflections of the silica matrix can be suppressed entirely by using contrast matching water. In such cases the surfactant aggregates in the pore can be studied against a uniform scattering contrast of silica matrix and water. The potential of these techniques for elucidating the self-assembly of surfactants in the pores will be demonstrated.

179c On the Fluctuation Theorem and Its Connection with Response Theory

Denis J. Evans, Research School of Chemistry, Australian National University, Canberra, Australia

Recently there has been considerable interest in the Fluctuation Theorem (FT). The FT shows how time reversible microscopic dynamics leads to irreversible macroscopic behavior as the system size or observation time increases. We show that the argument of the Evans-Searles FT, the dissipation function, plays a central role in nonlinear response theory and derive the Dissipation Theorem, giving exact relations for nonlinear response of classical N-body systems that are more widely applicable than any previous expressions. These expressions should be verifiable experimentally. When linearized they reduce to the well known Green-Kubo expressions for linear response.

Web Page: rsc.anu.edu.au/~evans/index.php

179d Quantum Sieving Effect Of Three-Dimensional Cu-Based Organic Framework For H2 And D2

Katsumi Kaneko, Chiba University, Faculty of Science, Chiba University, 1-33, Yayoi-cho, Inage-ku, Chiba-city, 263-8522, Japan, Chiba, 263-8522, Japan

The authors found remarkable soft nature of Cu-based organic framework on gas adsorption in the preceding report[1]. This paper describes new quantum molecular sieving effect of new type of Cu-based MOF. The crystal structure of [Cu(4,4'-bipyridine)2(CF3SO3)2]n metal-organic framework (CuBOTf) pre-evacuated at 383 K was determined with synchrotron X-ray powder diffraction measurement. Effective nanoporosity of the pre-evacuated CuBOTf was determined with N2 adsorption at 77 K. The experimental H2 and D2 adsorption isotherms of CuBOTf at 40 and 77 K were measured, being compared with quantum GCMC-simulated isotherms using the effective nanoporosity. The quantum-simulated H2 and D2 isotherms at 77 K using the Feynman-Hibbs effective potential coincided with the experimental ones. On the contrary, experimental H2 and D2 isotherms at 40 K had an explicit adsorption hysteresis, originating from the marked pore blocking effect on measuring the adsorption branch. The blocking effect for quantum H2 is more prominent than that for quantum D2; the selectivity for D2 over H2 at 40 K and 1 kPa was 5.8, being 1.5 times larger than that of single-walled carbon nanotube at least.

[1] Kondo, A; Noguchi, H; Ohnishi,S; Kajiro, H; Tohdoh, A; Hattori, Y; Xu, W.-C; Tanaka, H; Kanoh, H; Kaneko, K, Nano Lett. 2006, 6, 2581.

179e Multiscale Simulations Of The Thin Film Growth Of Organic Semiconductor Materials

Paulette Clancy, Cornell University, School of Chemical Engineering, Ithaca, NY 14853-5201

Semiconducting organic thin films are under intense scrutiny as materials for large area and mechanically flexible electronic displays. Among the small conjugated molecules suggested for organic electronic applications, pentacene is particularly important for its relatively high field effect mobility. While considerable progress has been made towards fabricating organic electronic displays, there is still much to be learned about the link between structure and molecular packing of pentacene thin films and the growth mechanisms that produce these structures.

We have undertaken multi-lengthscale simulations to understand how the performance (as measured by the order) of organic thin films is affected by the choice of organic molecule, substrate surface, and deposition conditions. The bonding interactions of pentacene molecules and thiophenes on a variety of surfaces (Si, oxide, metals) has been investigated using a combination of tight-binding simulations, ab initio Gaussian™ 03 calculations, kinetic Monte Carlo and Molecular Dynamics approaches. Each approach provides a different, and often complementary, piece of the puzzle to help understand how small organic semiconducting molecules deposit and grow into thin films with differing final properties. For instance, a fairly comprehensive set of Gaussian-derived adsorption energies allows the determination of preferred binding configurations as observed in experimental STM images. These same energies can also be used as the input to a mesoscopic scale study of the layer-by-layer deposition of organic thin films in lattice-based Kinetic Monte Carlo simulations to provide a rich and varied set of morphological "maps" that depend on a delicate balance between molecule-molecule and molecule-substrate interactions.

The way fluids flow into and fill nanopores is of interest to physicists, chemists, engineers and biologists. Nanoscale flow is dominated by surface properties and these can be studied directly using molecular simulation of model systems. Likewise the changes induced in biological systems by nanomaterials are relevant both to drug delivery and to toxicological response. In previous work we have considered equilibrium, steady state and transient flow in carbon nanopores as well as their response to dispersion in aqueous solutions. Recently we have extended our work to polymer surfaces such as PDMS including the effect of oxidation on aqueous flows. As a prelude to systematic studies of nanotoxicology we have also modeled the effect of carbon bucky balls on monolayers of the lung surfactant DPC.

In this lecture we review key results concerning flow in nanopores and membrane response to nanomaterials from theory, simulation and experiment. See for example:

S. Supple and N. Quirke, Rapid imbibition of fluids in carbon nanotubes, Physical Review Letters 90, 214501 (2003).

V. P. Sokhan, D Nicholson and N. Quirke, Transport properties of nitrogen in single walled carbon nanotubes J Chem Phys, 120, 3855 (2004).

S. Supple and N. Quirke, Nanocapillarity: II: Density profile and molecular structure for decane in carbon nanotubes, J Chem Phys 122, 104706 (2005)

M. Longhurst and N. Quirke, Environmental effects on the radial breathing modes of carbon nanotubes in water, J Chem Phys 124, 234708 (2006)

M. Longhurst and N. Quirke, Pressure dependence of the radial breathing mode of carbon nanotubes: The effect of fluid adsorption, Physical Review Letters 98, 145503 (2007)

M. Whitby and N Quirke, Fluid flow in carbon nanotubes and nanopipes, Nature Nanotechnology 2, 87 (2007)

302a Contact Angles, Pore Condensation And Hysteresis: Lessons From Simple Molecular Models

P. A. Monson, Univ. of Massachusetts, Dept. of Chem. Eng., Amherst, MA 01003

To be submitted.

302b Understanding the Dynamics of Proteins Domains

Coray M. Colina, Department of Materials Science and Engineering, Pennsylvania State University, 121 Steidle Building, University Park, PA 16802

To be submitted.

302c Observation Of Surface Phase Transitions At The Solid - Liquid Crystal Interface Via Molecular Simulation

Erich A. Muller[1], José M. Romero-Enrique[2], and Luis F. Rull[2]. (1) Chemical Engineering, Imperial College London, London, SW7 2AZ, United Kingdom, (2) Departamento de Física Atómica, Molecular y Nuclear, Universidad de Sevilla, Sevilla, Spain

Molecular simulations which evidence the existence of surface phase transitions of a Gay-Berne liquid crystal model vapor placed in contact with an unstructured attractive wall are reported. Monte Carlo simulations are performed in the Grand-canonical ensemble on a wide pore at a range of temperatures corresponding to states below and above the bulk nematic-isotropic-vapor triple point. The analysis of the adsorption isotherms show evidences of a first-order prewetting transition. In addition, a surface nematization of the first fluid layer close to the substrate, i.e. a surface phase transition between a disordered film and a nematic-like film, is observed. The nature of the ordered film is studied and is shown to be analogous to a two-dimensional Gay Berne fluid. For the lowest temperatures both transitions occur simultaneously. However, as the temperature is increased, surface nematization transition is observed after prewetting. At the highest temperatures, prewetting is not observed, and surface nematization occurs for isotropic liquid bulk conditions.

302d Computational Biology Using Order Parameters And Reaction Coordinates

Ravi Radhakrishnan, Department of Bioengineering, University of Pennsylvania, 120 Hayden Hall, 3320 Smithwalk, Philadelphia, PA 19104

Before I graduated from Keith's group (i.e., before he signed my thesis) Keith told me "if you are interested in an academic position, I have three words for you-- Multi Scale Modeling". When I gave my first talk on computational biology after I became a faculty member, Keith came up to me (rather disappointed) and said: "So you are doing Bio now, but atleast you get to do some statistical mechanics". So this talk is a little tribute to Keith on how we employ Statistical Mechanics and Multiscale Modeling to complicate complex biological systems even further.

I will present some recent results on how to utilize molecular dynamics, coarse-grained Monte Carlo, and Ginzburg-Landau simulations (things I learned in Keith's group) in understanding signaling through receptors in a cancer cell.

302e A New Material for the Selective Olefin/paraffin Separation by Adsorption

Lourdes F. Vega[1], Carmelo Herdes[2], Anabela Valente[3], Zhi Lin[3], Joao Rocha[3], and Joao A.P. Coutinho[3]. (1) MatGas Research Center. Air Products-CSIC-UAB, Campus de la UAB., Bellaterra. Barcelona, 08193, Spain, (2) Institut de Ciència de Materials de Barcelona, Consejo Superior de Investigaciones Científicas, ICMAB-CSIC. Campus de la UAB, Bellaterra. Barcelona, 08193, Spain, (3) Department of Chemistry, CICECO, Universidade de Aveiro, Aveiro, 3810-193, Portugal

The aim of the present work is twofold: (1) to propose an adsorbent material selective to ethyl chloride versus vinyl chloride and (2) to find a robust model, based on Monte Carlo simulations, for predictive purposes concerning the separation of ethyl chloride/vinyl chloride at optimal conditions. We present here results concerning the adsorption capacity of aluminum methyl phosphonate polymorph alpha (AlMePO-alpha) for pure ethyl chloride and vinyl chloride by measured/calculated individual adsorption isotherms of these pure compounds. Simulations were performed by the Grand Canonical Monte Carlo technique using our recently proposed molecular simulation framework for gas adsorption on AlMePO-alpha; We have improved an existing model for ethyl chloride in order to better capture the experimental dipole value; and equivalent molecular model for the vinyl chloride molecule was also developed for the simulation purposes. The molecular model of the material was used in a pure transferable manner from our previous work2. Simulations of the pure compounds were found to be in excellent agreement with the measured experimental data. Once the model was proved to be reliable, simulations were them used in a purely predictive manner as a tool to find the optimal conditions for the selective adsorption of these compounds. Results support the use of the AlMePO-alpha; as an appropriate adsorbent in the purification process of vinyl chloride, upholding the selective adsorption of the paraffin.

302f The Predictive Power Of Atomically-Detailed Simulations

Karl Johnson[1], Sudhakar V. Alapati[2], Bing Dai[3], and David S. Sholl[2]. (1) University of Pittsburgh, 1249 Benedum Hall, Department of Chemical Engineering, Pittsburgh, PA 15261, (2) Carnegie Mellon University, Department of Chemical Engineering, Carnegie Mellon University, Pittsburgh, PA 15213, (3) Chemical Engineering, University of Pittsburgh, 1249 Benedum Hall, Pittsburgh, PA 15261

Professor Gubbins has been one of the foremost advocates of the use of molecular simulations as a tool for predicting the behavior (thermodynamics, phase transitions, structure, etc.) of macroscopic systems. This talk focuses on the use of quantum mechanical methods for predicting the behavior of macroscopic systems, specifically considering the thermodynamics and kinetics of materials related to hydrogen storage in the solid state.

The on-board storage of hydrogen is one of the most vexing problems associated with the development of viable fuel cell vehicles. The challenge is to safely and efficiently store hydrogen at high volumetric and gravimetric densities. Hydrides of period 2 or 3 metals can store hydrogen at high gravimetric and volumetric densities. However, existing hydrides either have unacceptable thermodynamics or kinetics. New materials for hydrogen storage are therefore needed. We demonstrate how first principles density functional theory (DFT) can be used to screen potential candidate materials for hydrogen storage. We have used DFT calculations to screen over 300 reactions that have not been investigated experimentally. We have identified several interesting destabilized metal hydrides having favorable thermodynamics. We have predicted the van't Hoff plots for many of the promising candidate reactions identified though our modeling and as part of this work we have predicted van't Hoff plots for known reactions and compared our results with experimental data where available. We have also examined the effect of substitutional alloying on the thermodynamics of different metal hydrides.

We have studied the kinetics of selected surface reactions on several hydride-related materials using DFT. Our calculations help explain the difficulty of hydrogenating Mg_2Si, an otherwise promising destabilized metal hydride. Our calculations indicate that the initial dissociation of H_2 on Mg_2Si is facile at room temperature and above. However, our calculations also show that Mg_2Si is very susceptible to oxidation, which completely passivates the surface toward hydrogen dissociation.

302g Molecular Simulation Studies Of Protein-Surface Interactions

Shaoyi Jiang, Chemical Engineering, University of Washington, Benson Hall, Box 351750, Seattle, WA 98195

In this work, our recent efforts in the simulation of protein-surface interactions will be reviewed, particularly those related to the molecular-level nonfouling mechanism. Three approaches have been used in our molecular dynamics/Monte Carlo simulations to examine the origin of protein adsorption, including (a) surface hydration around varying surface functional groups, (b) repulsive forces applied to a protein when it approaches a surface, and (c) free energy changes upon protein adsorption. The surfaces studied include oligo(ethylene glycol) (OEG), phosphorylcholine (PC) and carbohydrate self-assembled monolayers. Simulations are performed using a locally developed simulation program, BIO_SURF, which is capable of simulating biomolecular-material interfaces in the presence of explicit water molecules and ions. Furthermore, quantum chemical calculations have been performed to examine the structures and charge distributions of a variety of zwitterionic groups in a vacuum or solvent. Accurate force fields for zwitterionic materials are developed based on ab initio results. With accurate force fields, molecular dynamics simulations are then performed to study how the packing structures of zwitterionic-based materials affect their ability to resist protein adsorption.

334a Application of Ewald Summations to Long-Range Dispersion Forces

Pieter J. in 't Veld[1], **Ahmed E. Ismail**[2], and Gary S. Grest[1]. (1) Surface and Interface Science Department, Sandia National Laboratories, PO Box 5800, MS 1415, Albuquerque, NM 87185, (2) Performance Assessment and Decisions Analysis Department, Sandia National Laboratories, PO Box 5800, MS 1395, Albuquerque, NM 87185-1395

We present results illustrating the effects of using explicit summation terms for the r^{-6} dispersion term on the interfacial properties of a Lennard-Jones fluid and SPC/E water. For the Lennard-Jones fluid, we find that the use of long-range summations, even with a short ``crossover radius,'' yields results that are consistent with simulations using large cutoff radii. Simulations of SPC/E water demonstrate that the long-range dispersion forces are of secondary importance to the Coulombic forces. In both cases, we find that the ratio of box size $L_{||}$ to crossover radius $r_c^{\mathbf{k}}$ plays an important role in determining the magnitude of the long-range dispersion correction, although its effect is secondary when Coulombic interactions are also present.

Sandia is a multiprogram laboratory operated by Sandia Corporation, a Lockheed Martin Company, for the United States Department of Energy under Contract No. DE-AC04-94AL85000.

334b Atomic-Continuum Simulation Of Flow Through "Smart" Porous Materials

Zhengmin Li and Donald Brenner. Materials Science and Engineering, North Carolina State University, Raleigh, NC 27695-7907

Stimuli-responsive polymer chains grafted into nanoporous materials can produce environment responsive fluid flow for applications that include smart drug delivery, bio-remediation and remote sensing. To accurately model these systems we are developing atomic-continuum fluid mechanics methods that couple flow dynamics with polymer chain structure and properties, which has been used previously for flow through polyglutamic acid, the Navier-Stokes-Brinkman equation is numerically solved using the structure and scaling properties for the polymer. Results will be compared to both direct molecular dynamics simulations for simple polymer models, and to experimental results for more complicated systems.

This work was supported by the National Science Foundation. Any opinions, findings and conclusions or recommendations expressed in this material are those of the authors and do not necessarily reflect the view of the NSF.

334c Computational Chemistry As A Guide For Understanding *f*-Element Chemistry In Alkaline Solutions

George S. Goff[1], Enrique R. Batista[2], Lia F. Brodnax[1], Michael R. Cisneros[1], Felicia L. Taw[1], and Wolfgang Runde[1]. (1) Chemistry Division, Los Alamos National Laboratory, Los Alamos, NM 87545, (2) Theoretical Division, Los Alamos National Laboratory, Los Alamos, NM 87545

Understanding the molecular level chemistry of *f*-element compounds (actinides (An) and lanthanides (Ln)) is critical for predicting behavior in complex solutions relevant to spent nuclear fuel reprocessing or nuclear waste repositories. While binary compounds (i.e., OH^- or CO_3^{2-} complexes) are fairly well understood, the understanding of *f*-element behavior in systems with multiple chelating ligands remains incomplete. Radiolysis products such as H_2O_2, formate, or oxalate are also strong complexing agents and can compete with carbonate and hydroxide to bind to the An and Ln ions, often forming multi-ligand solution species. The thermodynamic properties of these new complexes are often surprising (stability, solubility, etc.), and can be significantly different from the binary species. For example, H_2O_2 strongly complexes with U(VI) in solution to form $UO_2(O_2)(CO_3)_2^{4-}$, which is more than an order of magnitude more soluble than the previously thought solubility limiting species $UO_2(CO_3)_3^{4-}$ and is several orders of magnitude more thermodynamically stable.

Extraordinary advances have been made during the last two decades in quantum chemical studies of actinide hydroxo and carbonato complexes using DFT-based models, but the 5*f*-electronic structure still represents a challenge for quantum theoretical calculations. In this work we expand the application of computational chemistry to systems with multiple complexing ligands and show that hybrid DFT calculations show good agreement with experimental data (crystal structures, x-ray diffraction (XRD), UV-vis, Raman and IR vibrational frequencies) for a number of actinide solution complexes, such as $UO_2(CO_3)_3^{4-}$, $UO_2(O_2)(CO_3)_2^{4-}$, and $UO_2(O_2)_2(OH)_2^{4-}$. The remarkable agreement of experimental data with the calculated results gives good confidence in the models and allows us to extend the calculations to results that cannot be validated experimentally. For example, spectroscopic evidence (^{13}C-NMR and UV-vis) indicates the formation of a dimeric uranium carbonate compound with one peroxide, but this species has not been successfully isolated for single crystal XRD studies. DFT calculations have been successfully employed to show that the conformation of $(UO_2)_2(O_2)(CO_3)_4^{6-}$ with side-on coordinated peroxide is favored by 26 kcal/mol over an end on coordinated peroxide in aqueous solution.

334d Correlation-Based Coarse Graining Approach To Multiscale Materials Modeling

Carrie Y. Gao[1], Phani K. Nukala[2], **Bangwu Jiang**[1], Don M. Nicholson[2], David J. Keffer[1], and Brian J. Edwards[1]. (1) Chemical Engineering, University of Tennessee, Knoxville, TN 37996, (2) Computer Science and Mathematics Division, Oak Ridge National Laboratory

For systems with interactions that are well described by pair-wise potentials, the pair correlation function (PCF) provides a vehicle for passing information from the molecular level to the macroscopic level of description. It serves as natural starting point for multi-scale models that couple directly atomic regions embedded in coarse grained regions. In order to model a specific process related to mechanical properties, an effective Hamiltonian is required to govern dynamics. We propose to use the free energy as the effective Hamiltonian which consists of kinetic energy, potential energy and entropy. The free energy functionals of the PCF especially the entropy and the relationship to multiscale modeling are discussed. We describe the Hamiltonian first in the limit of a totally atomic system, second in the limit of an entirely coarse grained system and finally we discuss the coupling between the two descriptions. For pair wise interactions, the number density and pair correlation function are used for coarse graining.

334e Dft Investigation On The Stability Of Pt Clusters On Carbon Supports

Chethan Acharya and Heath Turner. Chemical and Biological Engineering, University of Alabama, Box 870203, Tuscaloosa, AL 35487-0203

A common anode catalyst used in a proton exchange membrane (PEM) fuel cell is platinum on a carbon support. The performance of the fuel cell can be enhanced by well-dispersed, stable platinum nanoparticles. One of the causes for the loss of activity in a fuel cell (as well as in other catalytic processes) is the growth, or sintering, of the nanoparticles. We suggest that the structural stability of nm-sized platinum catalyst particles can be enhanced by increasing the interaction of the nanoparticles with the carbon support, relative to the Pt-Pt interactions. In an effort to increase the interaction between the metal and the support, carbon substituted boron defects were introduced in graphite, and the adsorption energies of the metal clusters at absolute zero were estimated to be substantially higher when compared to pristine carbon using first-principles density functional theory (DFT) calculations. The adsorption energies of the metal clusters on oxidized and hydroxylated pristine and boron-doped graphite were also studied. The potential energy surface of the pristine and boron-doped carbon was scanned with a Pt atom and the activation barrier for its diffusion on the surfaces was estimated. The dynamics of Pt and Ru atoms on pristine and boron-doped graphite were studied at elevated temperatures with and without an external electric field using ab-initio molecular dynamics (AIMD) simulations and the mean square displacement of the metal atoms on boron-doped graphite were estimated to be lower than pristine graphite even at temperatures of 673 K. The adsorption energies of the gases prevalent at the anode side of the fuel cell such as CO and H2 was studied on pristine and boron-doped carbon supported Pt systems in addition to the CO and H2 oxidation kinetics. The activation barriers for these reactions were calculated.

334f Dft Study On The Kinetics Of Hydrogenolyses Of Xylene And Toluene

Yousef Sharifi, Chemical, Materials, and Biomolecular Engineering, University of Connecticut, 191 Auditorium Rd Unit 3222, Room 204, Storrs, CT 06268 and Luke E. K. Achenie, Department of Chemical Engineering, Virginia Polytechnic Institute and State University, Randolph Hall 133, Blacksburg, VA 24061.

For the last several years Carbon nanotubes have been at the center of attention because of their unique physical and chemical properties. One common method for the industrial production of carbon nanotubes is pyrolysis of hydrocarbons at high temperature. In the production of carbon nanotubes, quality and cost are very important factors; as such optimization of such reactors is very important. In order to carry out optimization studies, there is a need to know the chemical and physical phenomena inside the reactor. In this presentation we will discuss an Ab-Initio method for the study of gas phase pyrolysis of xylene and Toluene at high temperature in a CVD reactor. The structure of gaseous species were optimized using B3LYP/6-311(d,p) kevel of theory. Berny optimization is used to locate transition states structures. Our results are comparable to experimental data reported in the literature.

334g Exploring The Hyperthermal Deposition Of Pentacene Through Molecular Simulations

Joseph E. Goose and Paulette Clancy. Chemical and Biochemical Engineering, Cornell University, 713 E State Street, Apt 3, Ithaca, NY 14850

Pentacene is currently the standard bearer in the expanding field of organic semiconductors due to its relatively high electrical mobilities (rivalling amorphous silicon) and low production costs. One of the most successful methods of producing flat, ordered, thin films of pentacene is hyperthermal deposition. A simplistic view of hyperthermal deposition treats the deposition "event" or collision separately to longer timescale diffusional/thermal processes. The local effects of the collision are typically on the order of pico-seconds whereas thermal diffusion is a continuous process occurring over the lifetime of the film. In effect the deposition events are instantaneously altering the initial configuration in which the long term processes are taking place but not altering their intrinsic rates due to a rapid thermal equilibration.

We perform a statistical study of the collision event of an incident pentacene molecule in the proximity of a pentacene(010) step edge to relate molecular scale processes to experimentally observable data and features such as adsorption probabilities and AFM morphologies. Previous examples of this approach include a study of the trapping dynamics of ethane molecules on the Si(100) surface at comparable incident energies using empirical potentials.

We use the non-reactive empirical MM3 potential to model the collision of pentacene molecules in the proximity of a pentacene(010) step edge using molecular dynamics. The MM3 potential was used because it is known to accurately model organic molecules in both gaseous and crystalline forms and because of it's transferability to other organic molecules of interest. We show the process of hyper-thermal deposition of pentacene molecules to be sensitive to all variables studied, namely incident energy, incident angle, orientation upon impact and point of impact (in relation to a step edge) and quantified these effects. We have observed direct insertion events at energies of 1eV and above and the subsequent interstitials which are formed in the crystal. The sticking probabilities calculated are in good agreement with experimental values.

334h The Far-Infrared And Far-Raman Spectra Of Stable Metal Carbonyl Complexes At High Dilution. The Combination Of IR, Raman Spectroscopies And Density Functional Theory

Shuying Cheng[1], Martin Tjahjono[2], and **Marc Garland**[2]. (1) National University of Singapore, 4 Engineering Drive 4, Singapore, 117576, Singapore, (2) Process Science And Modelling, Institute of Chemical and Engineering Sciences, 1 Pesek Road, Jurong Island, Singapore, 627833, Singapore

Abstract: The Far-IR and Far-Raman spectra of [$Mn_2(CO)_{10}$], [$Mo(CO)_5$] and [$Re_2(CO)_{10}$] were investigated by both experiments and density functional theory (DFT) calculations. The present study consisted of three main parts. (1) Binary systems consisting of [$Mn_2(CO)_{10}$], [$Mo(CO)_5$] and [$Re_2(CO)_{10}$] in toluene under argon were experimentally measured in flow-through mode with far-IR spectroscopy using diamond windows. The BTEM algorithm, a spectral reconstructed method, was applied to these far-IR spectroscopic data to accurately reconstruct pure component spectra of [$Mn_2(CO)_{10}$], [$Mo(CO)_5$] and [$Re_2(CO)_{10}$]. (2) The same binary systems were measured by Raman spectroscopy in flow-through mode using a quartz cell. Again, the BTEM algorithm was applied to these Raman spectroscopic data to reconstruct pure component spectra of [$Mn_2(CO)_{10}$], [$Mo(CO)_5$] and [$Re_2(CO)_{10}$]. In above two experiments, the concentrations of organometallic systems under investigation are very dilute (ca. 100ppm) – similar to the concentrations typically used in metal-mediated homogeneous catalysis. (3) DFT calculations were performed, using different functionals and basis sets, for the optimized geometries of [$Mn_2(CO)_{10}$], [$Mo(CO)_5$] and [$Re_2(CO)_{10}$]. Then, both the far-IR and far-Raman spectra were predicted. The predicted far-IR and far-Raman spectra of [$Mn_2(CO)_{10}$], [$Mo(CO)_5$] and [$Re_2(CO)_{10}$] are rather consistent with the reconstructed IR and Raman spectra from experiments. In particular, the M-C vibrations in the IR and Raman could be assigned as well as the M-M vibrations in the Raman. The present results indicate that theoretical calculations combining with the experimental spectroscopies is a valuable tool to validate the vibrations of metal-metal, metal-carbon bonds of metal carbonyls at low wavenumber. Such a combined approach seems applicable to a wide range of structural identification problems in reactive organometallic systems, such as homogeneously catalyzed systems, in order to identify new species in solution.

Key words: Density functional theory, BTEM, Metal carbonyl, Far-Raman and Far-IR

334i The Hydration And Proton Dissociation In The Interaction Between Fluorocarbon-Based Phosphonic Acid And Water Clusters

Qin Liu[1], Oleg Borodin[2], and Grant D. Smith[2]. (1) Materials Science & Engineering, University of Utah, 122 S. Central Campus Dr. Rm 304, Salt Lake City, UT 84112, (2) Materials Science and Engineering, University of Utah, 122 S. Central Campus Dr. Rm 304, Salt Lake City, UT 84112

Here we present a theoretical study on the hydration and proton dissociation process in the interaction between CF3PO(OH)2 and water molecules. The hydration shell formation of CF3PO(OH)2 in complexes comprising n water molecules (n=6~12 for two CF3PO(OH)2 molecules) were studied using ab initio DFT methods . The starting configurations for the geometry optimizations were chosen from the most favorable positions. Further proton dissociation processes were observed when additional water molecules were added to the water clusters. The energy changes of the interactive system were also studied when hydronium and zundel ion forms during the proton transfer process. This study may help us understand the proton conduction process of the fluorocarbon-based polymers in low humidity conditions.

334j In Search Of Bilayer Undulations Via Molecular Dynamics Simulations And Principal Component Analysis

Allison N. Dickey and Roland Faller. Chemical Engineering and Materials Science, UC Davis, One Shields Ave, Davis, CA 95616

To determine how charge and lipid structure impact membrane stability, we use atomistic Molecular Dynamics simulations to study the undulatory behavior of large neutral and negatively charged lipid membranes. The results of an interesting theoretical study by Pincus et. al. found that lipid electrostatic interactions that are only partially screened by a weak electrolyte restrain bilayer undulations [1]. Besides density profile, lipid head group rotational correlation function, and order parameter comparisons, we examine global structural differences between the neutral and negatively charged bilayers using Principal Component Analysis (PCA). PCA is useful for analyzing high-dimensional data sets because it can be used to extract the dominant modes of a system. By applying PCA to lipid bilayers, we determine the relative magnitude of different fluctuation modes in the neutral and negatively charged bilayers, and through this comparison, enhance our understanding of the relationship between lipid electrostatics and experimentally obtainable parameters such as the lysis tension.

[1] Pincus, F.; Joanny, J.-F.;Andelman, D. Europhys. Lett. 1990, 11, 763.

334k Molecular Dynamics (Md) Study Of T To R Transition In 2ZN-Insulin Hexamer

Harish Vashisth and Cameron F. Abrams. Chemical Engineering, Drexel University, 3141 Chestnut St, Department of Chemical Engineering, Philadelphia, PA 19104

Insulin is a small (51 amino acids, 5800Da) polypeptide hormone that regulates carbohydrate metabolism, and is pharmacologically important in the treatment of type-1 diabetes mellitus. At high concentrations insulin aggregates to form hexamers which exist in one of the three allosteric states in solution, and are related by dynamic equilibrium: T6<=>T3R3<=>R6. Two kinds of important binding pockets exist in hexamers for various allosteric ligands: (1) Two ‘Zn binding pockets' which can bind single anions such as chloride, phenolate, carboxylates or thiocyanate (2) Six ‘hydrophobic binding pockets' located on dimer-dimer interfaces in hexamers which can bind small organic ligands such as phenol. Both kinds of allosteric ligands play a central role in T<=>R transition. Before binding to its transmembrane receptor, insulin must dissociate from hexameric storage form to the bioactive monomer. Hence, it is not difficult to infer that T<=>R equilibrium lies at the heart of insulin dissolution and anything affecting this rightly deserves scrutiny for pharmaceutical preparations of insulin. In this contribution, we use molecular simulations to investigate and understand this allosteric R to T transition in the presence of aforesaid ligands. We have carried out explicit-solvent, long (~20ns simulation time) molecular dynamics (MD) simulations of T-state insulin hexamers at two different concentrations of chloride as ligands to explore Zn binding domains. Our results show that Cl- ions bind more strongly as the concentration field is switched from 1.0 M to 1.5M, and are able to displace more water from Zn (II) pockets. This means, at increased concentration of Cl-, high-probability regions of binding are formed. Conformation of this stable binding by anions is an important step in understanding this transition. Further, in order to explore hydrophobic binding pockets, and establish importance of phenolic ligands, we carried out MD simulations again in explicit-solvent, both with and without phenol. Our hypothesis was that, as phenols stabilize R-state, a phenol free hexamer would make a transition to the T state. Indeed, we observed significant reduction in helical content of B1-B8 residues of each monomer in a ~20ns long MD simulation in NVT ensemble without all phenolic content. The melting of these helices is hallmark of this T<=>R transition. In order to strengthen this hypothesis and to eliminate the concern that the observed transition is due to thermostat or any other artifacts, we are presently carrying out long NVE simulations with explicit-solvent for the same cases as mentioned above. Interestingly, one of these ~10ns long run shows again the onset of melting of helices in three monomers out of six in the absence of phenols. This observation can be related to the important concept of co-operativity between subunits in all proteins. The work is underway to investigate this negative co-operativity and understand communication among subunits, and also to quantify residues involved in interaction of helix with phenols. The results of the present work will further serve to provide significant insights into structure-function relationship in insulin hormone and explain the mechanism of dissociation of insulin hexamer assembly to bioactive monomers.

334l Monte Carlo Simulation Of Porous Electrodes

Kenji Kiyohara, Research Institute for Cell Emgineering, National Institute of Advanced Industrial Science and Technology (AIST), 1-8-31 Midorigaoka, Ikeda, Osaka, Japan and Kinji Asaka, Reseach Institute for Cell Engineering, National Institute of Advanced Industrial Science and Technology (AIST), 1-8-31 Midorigaoka, Ikeda, Osaka, Japan.

We developed the Monte Carlo simulation technique in the constant voltage ensemble for the systems composed of electrodes and electrolytes. In this ensemble, the voltage between the anode and the cathode is specified as an external field rather than the surface charge density on the electrodes. We applied this technique to the systems with porous electrodes and different sizes of ions. We found that the asymmetry of the ion size causes the asymmetry in the structural, electrostatic, and electro-mechanical properties between the anode side and the cathode side. The effects of these physical properties on the electrical double layer capacitors and the electro-active polymers are discussed.

334m Prediction Of Binding Affinity From Ligand-Macromolecule Interaction Forces

Chandrika Mulakala and Yiannis Kaznessis. Chemical Engineering and Materials Science, University of Minnesota, 421 Washington Ave SE, Minneapolis, MN 55455

A major bottleneck in computational drug-design is the accurate computation of the binding affinity of a ligand-macromolecule interaction from first principles. When a series of analogous ligands with known affinities become available, quantitative structure-activity relationship (QSAR) development offers a viable alternative for binding affinity prediction. There, through statistical analysis of the structure-activity data-set, key physico-chemical characteristics of the ligands can be identified and their relative importance quantified, which can then be used for in silico screening of the next generation of drug leads.

However, the quantification of QSAR descriptors is ambiguous and often conformation dependent, and there is no universal descriptor set that can be applied to every system. There is therefore ample scope for the development of new QSAR descriptors to describe ligand-macromolecule interactions. We present here two new QSAR descriptors based on the interaction forces at the ligand-macromolecule binding interface, which together with the atomic mass of the ligand, correlate strongly with the binding affinity.

Our method is validated on two data-sets: compstatin analogs bound to C3, and fluorescein bound to antibody molecules. The two force-based descriptors, together with atomic masses of the ligands are apparently sufficient to capture both the enthalpic and entropic contributions to the free energy and therefore have great potential for binding affinity prediction. One limitation, however, is that these force-based descriptors require accurate ligand-macromolecule complex structures and will therefore need an x-ray crystallographic structure of at least one ligand-macromolecule complex.

334n Prediction of Mixture Vapor-Liquid Equilibrium from the Peng-Robinson Equation of State with Gex-Based Mixing Rule

Ming-Tsung Lee, Meng-Ting Hsieh, and Shiang-Tai Lin. Department of Chemical Engineering, National Taiwan University, No. 1, Sec. 4, Roosevelt Road, Taipei, 10617, Taiwan, Taipei, Taiwan

In this work we analyze the prediction of vapor-liquid equilibria (VLE) for a variety of binary mixtures over a wide range of temperature (183.15K – 623.15K) and pressure (0.1MPa -19 MPa) from the use of Peng–Robinson equation of state (PR EOS) and G^{ex}-based mixing rule, such as the Wong-Sandler (WS) and modified Huron-Vidal (MHV1) mixing rule. The predictive COSMO-SAC liquid activity coefficient model (LM) is used to provide the excess Gibbs free energy for the liquid phase.

In general, the predictions from the use of MHV1 mixing rule (average absolute error in pressure and in vapor phase composition are 2.5% and 4.9%, respectively) is more accurate than that from WS mixing rule (6.3%, 7.9%). It is found that the less accurate results from the WS mixing rule are owing to the inherent assumptions made. By means of analyzing excess energy, we found that the accuracy from the WS mixing rule can be improved with either of the two modifications: (1) The Stavermann – Guggenheim combinatorial term in the COSMO-SAC is ignored, and so the LM is denoted as $COSMOSAC^{res}$. The average error in both pressure and vapor phase composition from this approach is lowered to 4.0% and 4.7%, respectively. (2) Instead of matching G^{ex} from EOS and LM in the infinite pressure limit, we suggest to use the ambient condition instead. A detailed discussion for the improvements will be presented.

334o Reactive Molecular Dynamics Study For The Thermal Decomposition Of CF3OCF3

Bangwu Jiang, David J. Keffer, and Brian J. Edwards. Chemical Engineering, The University of Tennessee, 419 Dougherty Hall, Knoxville, TN 37996

Reactive Molecular Dynamics of the Thermal Decomposition of CF_3OCF_3

Bangwu Jiang, David J. Keffer, and Brian J. Edwards

Department of Chemical Engineering

University of Tennessee

Knoxville , TN 37996-2200 **Abstract**

In this work, we examine the feasibility of using conventional, well-tested, non-reactive interaction potentials for molecular dynamics of reactive systems undergoing chemical reaction. Our test system is the thermal decomposition of CF_3OCF_3, which is relevant as a simple molecule containing the necessary architectural elements to study chemical stability of perfluoropolyether lubricants. The reactivity of the Reactive Molecular Dynamics (RMD) procedure is implemented through the simulation algorithm, rather than through the potential. The molecular-level simulation algorithm must incorporate elements from both the quantum scale and the macroscopic scale.

Macroscopic Model: At the macroscopic scale, a chemical reaction that occurs via an elementary mechanism is completely defined by four specifications. The first specification is the stoichiometry of the chemical reaction. For the thermal decomposition of CF_3OCF_3, the stoichiometry is

(1)

Since this is an elementary reaction, we can write an expression for the macroscopic rate of reaction as

(2)

where k is a rate constant, and the square bracket represents concentrations. The rate constant can be expressed as

(3)

where k_o is a prefactor that represents the frequency at which the reaction is attempted, E_a is the activation energy of the prefactor, k_B is Boltzmann's constant, and T is absolute temperature. The reaction rate prefactor and the activation energy are the second and third specifications that must be made to uniquely describe the reaction at the macroscopic level. The fourth and final specification required is the heat of reaction, , which indicates the amount of heat generated or consumed by the reaction.

Quantum Mechanical Input: Quantum mechanical calculations provide information on the same four specifications, as required by the macroscopic model: stoichiometry, k_o, E_a, and . The thermal decomposition of CF_3OCF_3 was investigated by Pancansky *et. al*[1]. through quantum mechanics, in which they reported the three static properties: stoichiometry, E_a. and . We have repeated these calculations and additionally have calculated the rate constant, k_o, via transition state theory, using a combination of quantum mechanics and statistical mechanics.

Molecular Simulation Reaction Algorithm: We perform classical equilibrium molecular dynamics simulations in the canonical ensemble with a non-reactive potential. For the sake of this demonstration, we use the Universal Force Field[2]. The reactive algorithm has three steps. The first step is the reaction trigger, in which we check each CF_3OCF_3 molecule to determine if it is sufficiently near the transition state (both geometrically and energetically) to trigger a reaction. The determination of this state is based upon a mapping of the quantum mechanical transition state to a set of reaction geometric and energetic triggers. The geometric triggers ensure that the CF_3OCF_3 molecule is near the proper configuration, as determined through quantum mechanical calculations. There is an additional energetic trigger to determine if the relevant atoms have sufficient kinetic energy to overcome the known activation barrier. Each of these triggers has a numerical limit associated with it. The geometrical triggers are based upon quantum mechanical configurations. The energetic limit is tied to the known activation energy (from either quantum or experimental sources).

The second step of the reactive algorithm is the instantaneous reaction. Here we coarse-grain out the quantum mechanical details of the intermediate state of the molecule during chemical reaction. We replace the reactants (CF_3OCF_3) with the two product molecules (CF_4 and CF_2O), attempting to disturb the local structure as little as possible. The third and final step of this procedure is local equilibration, which is performed to satisfy exactly the target , which in turn ensures that, regardless of the reaction path taken, the final destination is correct. Once all three steps of the reactive algorithm are complete, we continue the classical simulation using the non-reactive potential.

Algorithm Validation: Using this procedure, we can match all four elements that uniquely define a macroscopic description of the reaction: the stoichiometry, k_o, E_a, and . The matching of the stoichiometry is mandated through the reaction trigger. The matching of the heat of reaction is mandated through the local equilibration following reaction. The matching of k_o and E_a are accomplished as follows. During an MD simulation, we monitor the rate of reaction. If we perform a suite of reactions varying only temperature, then we can generate an Arrhenius plot, which yields the k_o and E_a. We adjust the numerical values of the limits on the geometrical triggers to control k_o. We then adjust the numerical values of the limit on the energetic trigger to control E_a. The fitting of these parameters can be tuned to either quantum mechanical calculations or experimental measurements of k_o and E_a.

Algorithm Advantages: The RMD algorithm used here has several advantageous features. It is generalizable and has been used to model both the structural diffusion of protons and the thermal decomposition of perfluorinated ethers. The RMD algorithm has easy implementation because it fits in existing MD codes and is a computationally efficient multiscale scheme. The RMD algorithm can be quickly developed because it uses existing non-reactive interaction potentials and because most triggers are already known from TS without fitting. The RMD algorithm is rigorous, satisfying quantum mechanical transition state criteria and satisfying macroscopic models of reaction, with equal emphasis. The RMD algorithm is intrinsic adaptable; it can be applied to dilute or dense systems at high or low temperatures without reparameterization, because the triggers automatically capture the local environmental (in this case, density and temperature) dependence.

Acknowledgements:

This work has been supported by Air Force Office of Scientific Research through contract # FA 9550-05-1-0342. The authors wish to acknowledge resources of the Center for Computational Sciences at Oak Ridge National Laboratory, which is supported by the Office of Science of the DOE under Contract DE-AC05-00OR22725.

References:

(1) Pacansky, J.; Waltman, R. J. *Journal of Fluorine Chem.* **1997**, *82*, 79.

(2) Rappe, A. K.; Casevit, C. J.; Colwell, K. S.; Goddard, W. A.; Skiff, W. M. *J. Am. Chem. Soc.* **1992**, *114*, 10024.

334p Simulating Polymorphic Phase Behavior Using Reaction Ensemble Monte Carlo

John K. Brennan, Weapons and Materials Research Directorate, U.S. Army Research Laboratory, Building 4600, Aberdeen Proving Ground, MD 21005 and Martin Lisal, Institut for Chemical Process Fundamentals, Academy of Sciences of the Czech Republic, Rozvojova 135, Prague, CZ-165 02, Czech Republic.

We present an application of the reaction ensemble Monte Carlo (RxMC) method [J.K. Johnson, A.Z. Panagiotopoulos and K.E. Gubbins, Mol. Phys. 81, 717 (1994); W.R. Smith and B. Tøíska, J. Chem. Phys. 100, 3019 (1994)] for predicting solid-state structural transitions, namely conformational polymorphism. The RxMC technique provides a means of overcoming high-energy transition barriers which may be caused, for example, by sterically-hindered atoms or molecules. We demonstrate the method on nitromethane (NM), CH3NO2, whose behavior has been studied extensively. There are many morphologies of crystalline NM created by the rotation of the methyl group about the C-N bond. In applying the RxMC method to the phase transition behavior of crystalline NM, we treat the rotation of the methyl group as a conformational isomerization reaction. We then predict the equilibrium concentrations of the rotamers as a function of pressure and temperature. We find that the simpler, less-computationally expensive rigid model of nitromethane used in the RxMC simulation can capture the same behavior as the fully-flexible model. Investigations of such systems using the RxMC method allow for the identification of the key components of the mechanisms through the decoupling of the molecular degrees of freedom. In the scope of other similar solid-state transitions, the RxMC method can be a powerful tool in the interpretation of experimental measurements as well as in the confirmation of experimental findings.

334q Spatial Grand Canonical Monte Carlo Algorithms For Fluid Simulation

C. J. O'Keeffe, Ruichao Ren, and G. Orkoulas. Chemical and Biomolecular Engineering, University of California, 5531 Boelter Hall, 420 Westwood Plaza, Los Angeles, CA 90095

A new grand canonical Monte Carlo algorithm for continuum fluid models is proposed. The algorithm is characterized by the absence of strict detailed balance, and it is based on a generalization of sequential Monte Carlo algorithms for lattice gas systems. The elementary moves (particle insertions and removals) are constructed by analogy with a lattice gas system. The updating is implemented by selecting points in space (spatial updating) and the type of move (insertion or removal) is deduced based on the local environment of a point in space. Results on two-dimensional square-well fluids indicate that the proposed algorithm converges faster than all currently available grand canonical algorithms that satisfy strict detailed balance. Due to the nature of the updating, additional reduction of simulation time may be achieved by parallel implementation through domain decomposition. With strict detailed balance, parallel Monte Carlo simulation through domain decomposition cannot be validated with conventional Markov chain theory, which describes an intrinsically serial stochastic process. We show that spatial updating is the key to improve efficiency in parallel simulations through domain decomposition. A parallel scheme is proposed to reduce inter-processor communication or synchronization which slows down parallel simulation with increasing number of processors. Our parallel simulation results show substantial reduction of simulation time for systems of moderate and large size.

Thermodynamic Stability of Hydrogen Storage Material $Zn(BH_4)_2$: First-Principles Study

Pabitra Choudhury[1, 2], Venkat R. Bhethanabotla[,1, 2] and Elias Stefanakos[2]*
[1]Sensors Research Laboratory, Department of Chemical Engineering
[2]Clean Energy Research Center, Department of Electrical Engineering
University of South Florida, 4202 E. Fowler Ave., ENB 118, Tampa, FL 33620

Abstract

Ab initio calculations using density functional theory were performed to establish the most stable crystal structure of $Zn(BH_4)_2$. It was found to be of orthorhombic structure in the space group of $Pmc2_1$ with lattice parameters of a = 4.118 Å, b = 4.864 Å, c = 7.916 Å. It is an insulating material having a band gap of around 3.529 eV. Our calculations showed nearly tetrahedral shape of the BH_4 complex with B-H bond lengths $d_{B\text{-}H}$ = 1.20-1.25 Å and H-B-H bond angles $\theta_{H\text{-}B\text{-}H}$ = 104.16-120°. The reaction enthalpy was also found for the reaction $Zn(BH_4)_2 = Zn + 2B + 4H_2(g)$ to be 59.31 kJ/mol of H_2 including zero point energy at 0 K.

Introduction

In hydrogen storage materials, one recent challenge is to find light-weight complex solid hydrides which have high gravimetric theoretical capacity exceeding the 2010 DOE target of 6.5 wt % for on-board vehicular application. The technical challenge is to find materials that can exhibit favorable thermodynamics and kinetics for hydrogen de-sorption and absorption, and have ability to store a sufficient amount of hydrogen by weight as well as by volume percent measures. Boro-hydride complexes as hydrogen storage materials have recently raised great interest. Most of the commonly known boro-hydrides are found to be unsuitable for on-board application because of either high stability (resulting from a very high decomposition temperature) or complete irreversibility[1]. $Zn(BH_4)_2$ is considered to be a potential candidate for on-board application as it has high theoretical hydrogen storage capacity of 8.4 wt% and also a reasonably low decomposition temperature (85 °C)[2] compared to complex alkali borohydrides (which decompose above 300 °C).

A combined theoretical and experimental approach can better lead us to design a suitable complex material for hydrogen storage. Here, we explore the complex hydrides using density functional theory (DFT) to provide information at

[*] Author for correspondence at venkat@eng.usf.edu

the electronic level that is hard to obtain by experimental methods. Such calculations can also unravel the transition states of the molecules undergoing chemical transformation at complex solid surfaces. Phonon spectrum helps us to understand the finite temperature crystal stability and such calculations were also performed in this work.

Computational details and methodology

First-principle calculations were performed in this work on $Zn(BH_4)_2$ using DFT within the local density approximation (LDA) and projector augmented wave (PAW) method and a plane wave basis set to calculate the total energies, as implemented in the Vienna *Ab initio* Simulation Package (VASP).[3-5] We started our calculations from known similar complex formula units $X(YZ_4)_2$ (X=Be, Ca, Mg and Cu; Y=Al and B; Z=H, Cl, Br and I) for $Zn(BH_4)_2$ to find the most stable crystal structure. We have optimized the atomic positions and cell parameters for each model and finally optimized the cell volume against the total energy of the system. We have calculated the enthalpy for the formation equation of $Zn(BH_4)_2$ from its elemental components as given below:

$$Zn + 2B + 4H_2(g) \rightarrow Zn(BH_4)_2 \quad (1)$$

The calculated results indicated that those for space group *Pmc2₁* give the highest negative enthalpy of formation for $Zn(BH_4)_2$ which was considered the most stable structure. We also calculated the electronic electron localization function (ELF) of $Zn(BH_4)_2$ to gain insights into the nature of the material and bonding between the elements. ELF is calculated from the total electron density, its gradient and the kinetic energy density. It is associated with the probability density of finding two electrons in the same spin-state close to each other.

Based on the optimized crystal structure from DFT, lattice dynamics based on the harmonic approximation using the direct force constant method in the PHONON[4] code was used to calculate the phonon density of states (DOS) to confirm the finite temperature crystal stability using the positive definite frequency distribution criteria over the Brillouin zone.

Results/Discussion

The above methods were used to establish the stable crystal structure with appropriate lattice parameters & atomic positions and the electronic structure of the high theoretical hydrogen storage capacity material $Zn(BH_4)_2$. The minimum total energy for $Zn(BH_4)_2$ was found to be -44.595 eV/mol at a unit cell volume of 150.82 $Å^3$. The final optimized crystal structure with atomic positions is shown in Table 1.

Table 1: Optimized crystal structure of $Zn(BH_4)_2$. The space group is *Pmc2₁* (#26) with the lattice parameters a = 4.118 Å, b = 4.864 Å, c = 7.916 Å. All the atomic positions are given by Wyckoff letter and the number of formula units in the unit cell Z=2.

Element	Wyckoff letter	X	Y	Z
Zn	2*a*	0.00000	0.28459	0.00890
B1	2*b*	0.50000	-0.06481	0.47060
B2	2*a*	0.00000	-0.46194	0.24842
H1	2*b*	0.50000	-0.30735	0.42745
H2	4*c*	0.27449	-0.00500	-0.42983
H3	2*a*	0.00000	0.39706	0.37973
H4	2*a*	0.00000	0.21066	-0.24017
H5	4*c*	0.25820	-0.48360	-0.32151
H6	2*b*	0.50000	0.08396	0.34880

The calculations indicated that $Zn(BH_4)_2$ is orthorhombic structure having lattice constants a = 4.118 Å, b = 4.864 Å, c = 7.916 Å and nearly tetrahedral in shape for the BH_4 complex with B-H bond lengths $d_{B\text{-}H}$ = 1.20-1.25 Å and H-B-H bond angles $\theta_{H\text{-}B\text{-}H}$ = 104.16-120°. The ELF (100 plane) calculated here (Fig. 1) shows strong and large attractors around the H atoms. As hydrogen has no core attractor, this indicates either a shared-electron or a closed-shell bond. The very low ELF value on the B sites indicates delocalized electrons, while a spherical shell attractor is seen around Zn. The very low ELF value between Zn and $[BH_4]^-$ complex indicates a low ionic bond, whereas, a very high value of ELF within the $[BH_4]^-$ complex indicates iono-covalent bonding.

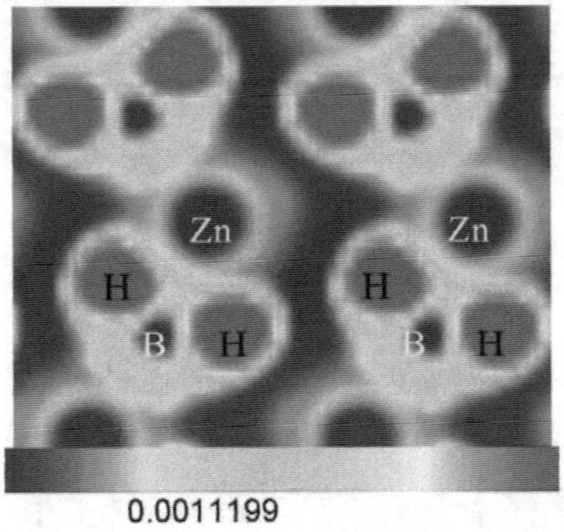

Figure 1. Electron localization function (ELF) of $Zn(BH_4)_2$ (100 plane)

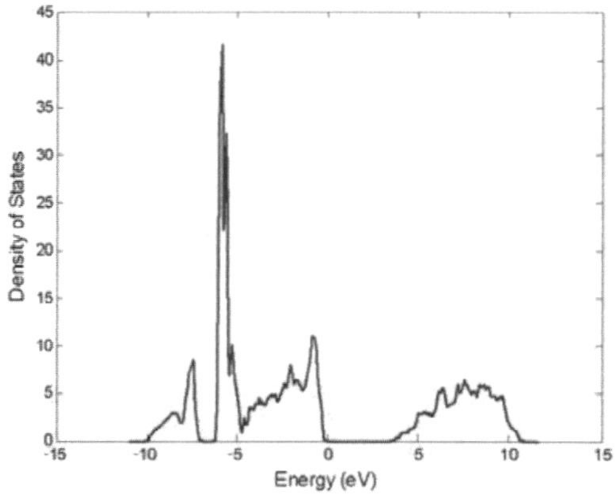

Figure 2. DFT/LDA electronic total density of states (DOS) relative to Fermi level

The total density of states (DOS) shows that there is a wide band gap between the total valence band and the conduction band which confirms the insulating properties of the $Zn(BH_4)_2$. It also shows that the valence band has three parts for $Zn(BH_4)_2$, whereas, the alkali and alkaline boro-hydrides have two major parts in the valence band. This may be due to the presence of static *d* electrons from the Zn atom. However further investigations are required before providing more specific interpretations.

Conclusions

The $Zn(BH_4)_2$ is found to be orthorhombic structure in the space group *Pmc2₁*. The electronic structure analysis shows $Zn(BH_4)_2$ is an insulating material having a band gap of 3.529 eV and strong covalent bonding between hydrogen atoms and boron in $[BH_4]^-$ complex and also less polar bonding between Zn and hydrogen atom. The enthalpy of formation of -59.31 kJ/mol of H_2 including ZPE at 0 K suggests that the crystal structure is stable at room temperature and thus $Zn(BH_4)_2$ can be considered a potential candidate for reversible hydrogen storage.

Reference

1. W. Grochala and P. P. Edwards, Chem. Rev. **104**, 1283 (2004).
2. E. Jeon and Y. Cho, Journal of Alloys and Compounds (2006).
3. G. Kresse and J. Furthmüller, Phys. Rev. B **54**, 11169 (1996).
4. G. Kresse and J. Furthmüller, Comput. Mater. Sci. **6**, 15 (1996).
5. G. Kresse and J. Hafner, Phys. Rev. B **47**, R558 (1993).
6. K. Parlinski, Software PHONON (2005)

334s Thermophysical Properties Of Aromatic Compounds: Coupling Quantum Mechanics And Statistical Thermodynamics

Mohamad H. Kassaee, University of Tennessee, 1700 W. Clinch Ave, Apt 505, Knoxville, TN 37916, **Jared T. Fern**, Physical Properties Research Facility, The University of Tennessee, 327 Dougherty Engineering Bldg., 1512 Middle Drive, University of Tennessee, Knoxville, Knoxville, TN 37831-6273, Benjamin M. Hoffman, Chemical Engineering, The University of Tennessee, Knoxville, TN 37996-2200, David J. Keffer, The University of Tennessee, 617 Dougherty Hall Dougherty Hall, Knoxville, TN 37996-2200, and William V. Steele, ORNL/Chemical Engineering, University of Tennessee, Oak Ridge National Laboratory, Oak Ridge, TN 37831-6181.

In this paper, we examine the practical procedure used to obtain the thermophysical properties of various compounds in the ideal gas reference state [1-2]. This procedure uses both quantum mechanics and statistical mechanics to calculate the thermophysical properties. Using *ab initio* calculations, the electronic structure of the molecule is optimized. Normal vibrational frequencies, energetic barriers to internal rotations and moments of inertia are calculated from the optimized structures using quantum mechanics. Given the current state-of-the-art numerical solutions of the Schrödinger equation, it is necessary to apply an empirical scaling factor to the normal vibrational frequencies computed by *ab initio* means [3]. Averaged scaling factors over samples of molecules are tabulated for a variety of quantum methods and basis sets [4]. In this paper, we use averaged scale factor over simple aromatic compounds since the studied compounds are all aromatic.

Various contributions to entropy are calculated. The internal rotation modes are distinguished from the vibrational modes by creating movies from each mode's eigenvector. We calculate the relaxed energy barrier for each rotor using quantum mechanics. Taking the empirically scaled frequencies and internal rotations from quantum mechanical calculations, statistical mechanics is used to calculate translational, vibrational, rotational and internal rotation contributions to the entropy. Relative enthalpies are calculated from quantum mechanics and coupling it with entropies yields most of the important thermophysical properties.

We present an evaluation of the combined quantum mechanical and statistical mechanical procedure for generating reference entropies, including refinements from previous work. Additionally, we present new data for phenathrene, methyl phenanthrenes, and dimethyl phenathrenes. Our standard, by which the procedure is evaluated, is a set of highly precise experimental measurements of the reference entropies for these compounds.

Acknowledgements

The authors gratefully acknowledge the financial support of the Office of Fossil Energy of the U.S. Department of Energy (DOE). Through the UT Computational Science Initiative, this research project used resources of the Center for Computational Sciences at Oak Ridge National Laboratory, which is supported by the Office of Science of the DOE under Contract DE-AC05-00OR22725.

References

1. Kassaee, M. H.; Keffer, D. J.; Steele, W. V., *A comparison between entropies of aromatic compounds from quantum mechanical calculations and experiment.* Journal of Molecular Structure (Theochem) **2007**, 802, 23-34.

2. Kassaee, M. H.; Keffer, D. J.; Steele, W. V., *Theoretical Calculation of Thermodynamic Properties of Naphthalene, Methylnaphthalenes and Dimethylnaphthalenes.* submitted to J. Chem. Eng. Data, **04/2007**.

3. Scott, A. P.; Radom, L., *Harmonic Vibrational Frequencies: An Evaluation of Hartree-Fock, Moller-Plesset, Quadratic Configuration Interaction, Density Functional Theory, and Semiempirical Scale Factors.* Journal of Physical Chemistry, **1996** 100(41), 16502-16513.

4. NIST, *Computational Chemistry Comparison and Benchmark DataBase.* **2007**, National Institute of Standards and Technology.

377a Replica Exchange for Reactive Monte Carlo Simulations

Heath Turner, Chemical and Biological Engineering, University of Alabama, Box 870203, Tuscaloosa, AL 35487-0203, John K. Brennan, Weapons and Materials Research Directorate, U.S. Army Research Laboratory, Building 4600, Aberdeen Proving Ground, MD 21005, and Martin Lisal, Institut for Chemical Process Fundamentals, Academy of Sciences of the Czech Republic, Rozvojova 135, Prague, CZ-165 02, Czech Republic.

A new application of the replica exchange technique is demonstrated for use in reactive Monte Carlo simulations. The technique allows the system to traverse barriers in the free energy landscape through configuration swaps of higher temperature replicas, replicas with scaled potential energy parameters, or replicas existing at modified values of the reaction driving force (free energy of reaction). Acceptance rules are developed for the swap moves among the replicas, and the application of the replica exchange method is shown to be straightforward. We demonstrate the accuracy and efficiency of this technique by simulating the conversion of a model reaction. We find large increases in efficiency when performing simulations at computationally challenging conditions, such as high densities, particularly when a pre-converged configuration reservoir is incorporated. With minimal effort, it should be possible to extend this replica exchange approach to systems involving multiple simultaneous reactions and to systems with more complex multi-site reacting species.

377b Reaction Coordinates And Transition Pathways Of Rare Events Via Forward Flux Sampling

Ernesto E. Borrero-Quintana, Chemical and Biomolecular Engineering, Cornell University, 120 Olin Hall, Ithaca, NY 14853 and Fernando A. Escobedo, School of Chemical and Biomolecular Engineering, Cornell University, 120 Olin Hall, Ithaca, NY 14853.

Path sampling methods based on importance sampling of dynamical trajectories have become powerful simulation tools for understanding dynamics in complex systems. However, the challenges of applying these path sampling algorithms, such as Forward Flux Sampling (FFS), to rare events include: (i) determination of an adequate order parameter (or combination of variables) that allows the description of multiple transition state regions of a process; (ii) assessing efficiency and completeness of sampling; and (iii) estimation of the free energy profile and barriers. These challenges arise, in large part, because the efficiency of these sampling schemes is sensitive to an improper choice of order parameter. To address these limitations, we present a new approach for identifying suitable reaction coordinates to describe the progression of rare events in complex systems. The method is based on a FFS-type sampling technique and standard least square estimation (LSE) and it is therefore denoted as FFS-LSE. The algorithm generates trajectories for the transition between stable states as chains of partially connected paths, which can be then used to obtain "on-the-fly" estimates for the committor probability to the final region "p_B". These p_B data are then used to screen a set of candidate collective properties for an optimal order parameter (i.e., reaction coordinate) that depends on a few relevant variables. LSE is then used to find the coefficients of the proposed reaction coordinate model, and an analysis of variance is used to determine the significant terms in the model. We also introduce a new method to evaluate simultaneously the kinetics and the free energy profile in one single FFS simulation series. The approach computes free energies using the Monte Carlo ensemble of trajectories generated by the FFS-type sampling scheme combined with a reweighing procedure that resembles an umbrella sampling between interfaces. These approaches are demonstrated for several problems including protein folding of both simple lattice models as well as atomistic models like that for the ß-sheet Forming Binding Protein (FBP28) WW domain. We showed that a simple approximation to pB as order parameter in terms of collective variables is sufficient for describe the intrinsic dynamics of the complex system and ensure a complete and efficient sampling. Furthermore, since the p_B surface found from the FFS-LSE approach leads to the identification of the free energy profile in terms of the estimated reaction coordinate, the ensemble of transition states and mechanistic details of the transition can be readily obtained without the need of additional simulations.

377c A New Paradigm In Force Dependence In Single Molecule Experiments Unraveled By Molecular Simulations

Ravi Radhakrishnan and Ravindra Venkatramani. Department of Bioengineering, University of Pennsylvania, 120 Hayden Hall, 3320 Smithwalk, Philadelphia, PA 19104

High fidelity polymerases are efficient catalysts of phosphodiester bond formation during DNA replication/repair. We interpret molecular dynamics simulations of a polymerase bound to its substrate DNA and incoming nucleotide using a quasi-harmonic model to formulate a new model for studying the effect of external forces applied to the bound DNA on the kinetics of phosphoryl transfer. The origin of the force dependence is shown to be an intriguing coupling between slow, delocalized polymerase-DNA modes and fast catalytic site motions. Using non-cognate DNA substrates we show that the force dependence is context specific. We validate the simplying assumptions of our quasiharmonic model by performing mixed QMMM simulations of the polymerase catalysis reaction.

377d Long-Range Lennard-Jones And Electrostatic Interactions In Interfaces: Application And Development Of The Isotropic Periodic Sum Method

Jeffery Klauda[1], Xiongwu Wu[2], Richard W. Pastor[2], and Bernard R. Brooks[2]. (1) Department of Chemical and Biomolecular Engineering, University of Maryland, College Park, MD 20742-2111, (2) Laboratory of Computational Biology, National Heart, Lung, and Blood Institute, National Institutes of Health, 50 South Dr., Bethesda, MD 20892

Molecular dynamics (MD) simulations of heptane/vapor, hexadecane/vapor, water/vapor, hexadecane/water, and dipalmitoyl phosphatidylcholine (DPPC) bilayers and monolayers are analyzed to determine the accuracy of treating long-range interactions in interfaces with the isotropic periodic sum (IPS) method. Method and cutoff (r_c) dependence of surface tensions, density profiles, water dipole orientation, and the electrostatic potential profiles are used as metrics. The water/vapor, heptane/vapor, and hexadecane/vapor interfaces are accurately and efficiently calculated with 2D IPS (r_c=10 Å). It is demonstrated that 3D IPS is not practical for any of the interfacial systems studied. However, the hybrid method PME/IPS (Particle Mesh Ewald for electrostatics and 3D IPS for Lennard-Jones (LJ) interactions) provides an accurate and efficient way to include both types of long-range forces in simulations of large liquid/vacuum and all liquid/liquid interfaces, including lipid monolayers and bilayers. For liquid/vacuum interfaces, a more efficient 2D+1D IPS method is developed with no loss of accuracy. The contribution to surface tension of LJ terms arising from interactions beyond 10 Å range from 13 dyn/cm for the hexadecane/vapor interface to approximately 3 dyn/cm for hexadecane/water, and DPPC bilayers and monolayers. Surface tensions of alkane/vapor, hexadecane/water, and DPPC monolayers based on the CHARMM lipid force fields agree very well with experiment, while surface tensions of the TIP3P and TIP4P-Ew water models underestimate experiment by 16 and 11 dyn/cm, respectively. Dipole potential drops ($\Delta\Psi$)•are less sensitive to long-range LJ interactions than surface tensions. However, $\Delta\Psi$ for the DPPC bilayer (845±3 mV proceeding from water to lipid) and water (547 ±2 mV for TIP4P-Ew and 521±3 mV for TIP3P) overestimate experiment by factors of 3 and 5, respectively, and represent expected deficiencies in non-polarizable force fields.

377e Continuous Fractional Component Monte Carlo: The General Biased Slow-Growth Method For Phase Equilibrium Simulations

Wei Shi, Chemical and Biomolecular Engineering, Univ. of Notre Dame, Notre Dame, IN 46556 and Edward J. Maginn, Department of Chemical and Biomolecular Engineering, Univ. of Notre Dame, Notre Dame, IN 46556.

We have recently developed an open system Monte Carlo procedure designed to overcome difficulties with insertion and deletion of molecules [1], particularly for systems involving complex molecules having specific interactions. The method utilizes gradual insertions and deletions of molecules through the use of a continuous coupling parameter and an adaptive bias potential. The method draws upon concepts from previous open system molecular dynamics and expanded ensemble Monte Carlo techniques, and we show how it can be applied within the grand canonical, Gibbs and osmotic ensembles. It is shown to yield correct results for the volumetric properties of the Lennard-Jones fluid and water as well as the phase behavior of the CO2- ethanol binary system. It is then used to compute absorption isotherms, excess molar volumes and enthalpies of absorption for CO2 and water in the ionic liquid 1-n-hexyl-3-methylimidazolium bis(trifluoromethylsulfonyl)imide. The results are in quantitative agreement with experimental data.

[1] Wei Shi and Edward J. Maginn, J. Chem. Theory Comput. 2007, in press.

377f Understanding Polymorph Selection during Crystal Nucleation and Growth

Jerome P. Delhommelle, Department of Chemical Engineering, University of South Carolina, 301 South Main Street, Columbia, SC 29208 and Caroline Desgranges, Chemical Engineering, University of South Carolina, 301 South Main Street, Columbia, SC 29208.

Controlling the structure (or polymorph) in which a molecule crystallizes is a long-standing issue. Since polymorphs have different physical properties, it is crucial for many applications to control which polymorphs form during crystallization. We develop hybrid Monte Carlo methods and apply them to study polymorph selection during crystal nucleation and growth. We perform two different types of simulation methods corresponding to the two steps of nucleation and growth. Crystal nucleation is an activated process. Therefore, we combine these hybrid methods with a non-Boltzmann sampling technique (the so-called umbrella sampling technique) and an appropriate reaction coordinate. Once we have formed a crystal nucleus of a critical size, we carry out hybrid simulations of the growth from the crystal nucleus without applying any bias potential. We illustrate these methods on a series of examples.

377g Statistically Optimal Free Energy Estimates without Histograms

Michael R. Shirts, Department of Chemistry, Columbia University, Havemeyer Hall MC 3131, New York, NY 10027 and John D. Chodera, Department of Chemistry, Stanford University, Chemistry Operations MC 5080, Stanford, CA 94605-5080.

We present a maximum-likelihood method to combine data from multiple equilibrium simulations conducted at different thermodynamic states, based on statistical methods for correcting for biases in sampling [1]. The resulting estimator of the free energies yields the minimum-variance asymptotically unbiased estimate of the true free energy. This estimator is equivalent to multiple histogram method [2] or weighted histogram analysis method method [3] in the limit where the histogram bins are taken to be vanishingly small, but is derived without the need to invoke histograms, thus eliminating bias caused by binning. This is especially useful for highly multidimensional problems where histograms are difficult to populate.

Additionally, this method provides an estimator to directly compute uncertainties in the relative free energies from the data, something not possible with previous methods. This methodology is expected to be widely useful in the analysis of "alchemical" free energy calculations, generalized ensemble simulations, and the calculation of potentials of mean force.

[1] D. Gill, Y. Vardi, and J. A. Wellner , Annals of Stat. 16(3):1069-1112 (1988)

[2] A. M. Ferrenberg and R. H. Swendsen, Phys. Rev. Lett. 63:1195-1198 (1989)

[3] S. Kumar, D. Bouzida, R. H. Swendsen, P. A. Kollman and J. M. J. Rosenberg, J. Comput. Chem., 13, 1011-1021 (1992)

Web Page: www.columbia.edu/~mrs2144/MRSCV.html

377h An Evaluation of Learning-Augmented Monte Carlo

Eric Jankowski and Sharon C. Glotzer. Chemical Engineering, University of Michigan, Ann Arbor, MI 48109

A common yet difficult problem encountered in molecular simulations is given a set of N molecules and their interactions, what configuration minimizes their potential energy? In 2005, Troisi, Ratner and Wong developed a novel algorithm for finding energy minimizing configurations of particles. Rigid, lattice-based molecules are able to learn about local clusters of molecules that minimize potential energy and bias the formation of these clusters. By moving these clusters as whole units, the algorithm finds configurations of lower potential energy in smaller amounts of time than a traditional Metropolis Monte Carlo simulation. We improve upon the Troisi algorithm and quantify its performance gains vs. Monte Carlo with cluster moves, a more meaningful comparison. We find that the learning algorithm does not sample the NVT ensemble and performs no better than cluster Monte Carlo in finding energy-minimizing configurations.

Jankowski, E. & Glotzer, S.C. (2007) *Preprint.*

Troisi, A., Ratner, M. & Wong, V., (2005) *Proc. Natl. Acad. Sci. USA* **102** 255-260.

377i Free Energy Calculations In Field Theoretic Simulations

Erin Lennon, Chemical Engineering, University of California, Santa Barbara, Engr. II, Room 3357, UCSB, Santa Barbara, CA 93106-5080, Kirill Katsov, Materials Research Laboratory, University of California, Santa Barbara, Engr. II, Room 3357, UCSB, Santa Barbara, CA 93106-5080, and Glenn H. Fredrickson, Chemical Engineering and Materials, University of California, Santa Barbara, Engr. II, Room 3357, UCSB, Santa Barbara, CA 93106-5080.

Free energy calculations are well-established in particle based simulations, but similar tools have yet to be fully developed for systems described by continuous fields discretized on a spatial lattice. These field-based systems are of particular importance when studying macromolecules and their self-assembly. For example, the functionality of polymeric materials is strongly correlated with their mesoscale equilibrium structure. The study of systems on this scale is often intractable by traditional particle based MD and MC simulations. Previous work has shown how field theoretic simulations can be used to appropriately investigate these systems in the mean-field limit (for a review, see [1]); however, progress beyond this approximation is limited by a lack of techniques for determining the free energy of field-based systems. In this work, we present a novel application of traditional thermodynamic integration methods to discretized field theoretic models. Using this to determine the absolute free energy of a diblock copolymer system, we qualitatively study the effect of polymer size on the order-disorder transition. [1] G.H. Fredrickson, *The Equilibrium Theory of Inhomogenous Polymers* (Clarendon Press, 2005).

377j A Novel Algorithm For Creating Density Dependent Implicit Solvent Models

Erik C. Allen and Gregory C. Rutledge. Chemical Engineering, Massachusetts Institute of Technology, 38 Bedford Street, Quincy, MA 02169

Implicit solvent simulations, i.e. those in which solvent molecules are not explicitly simulated and the solute-solute potential is modified to incorporate this solvent effect, have the potential to speed up many relevant molecular simulations, including studies of proteins and surfactants. While a pairwise potential between solute sites is simplest, the additional parameterization based on the local solute concentration has the potential to increase the accuracy of the simulations with only a marginal increase in computational cost. We describe here a method in which the radial distribution function and free energy of solvation of a system of interest are first measured in a fully explicit solvent simulation, and then reproduced across a range of densities in an implicit solvent simulation.

432a A Conservative Algorithm For An On-The-Fly Change Of Resolution In Mixed Atomistic / Coarse-Grained Multiscale Simulations

Andreas Heyden and Donald G. Truhlar. Chemistry, University of Minnesota, 228 Smith Hall, 207 Pleasant Street S.E., Minneapolis, MN 55455

Coarse-graining techniques are commonly used to increase the computational efficiency of molecular simulations of complex soft matter systems. A difficulty in applying these methods for describing complex systems is the need to obtain accurate coarse-grained potential energy parameters for strongly interacting systems where atoms become polarized or react. As a result, coarse-graining techniques often cannot be applied to these systems, and less efficient but generally more accurate all-atom force fields have to be employed. In order to combine the efficiency of coarse-grained force fields with the accuracy of atomistic methods, mixed-resolution force fields have been developed that use a coarse-grained description for the part of the system distant from an active site and an atomistic description for the active site and its direct environment. Since atoms or groups of atoms may diffuse during a molecular simulation, the algorithm needs to permit an on-the-fly reclassification (from atomistic to coarse-grained or vice versa) of atoms or groups of atoms as they transition between the high- and low-resolution regimes. Previous algorithms for adaptive-resolution molecular dynamics simulations are equivalent to using non-conservative force fields, which result in a lack of conservation of both total system energy and total angular momentum.(1-3)

Recently, we introduced an adaptive partitioning method for treating the potential energy function for systems with an active zone modeled at a high-level of theory and a surrounding environmental zone modeled at a low-level of theory.(4) This algorithm permits an on-the-fly change in level of theory of the potential energy function of an atomistic simulation as atoms or groups of atoms transition between the active and environmental zones. By extending this adaptive partitioning method for potential energy functions to adaptive partition Lagrangian functions, we were able to derive a Hamiltonian for mixed-resolution systems that allows for a change in resolution of selected groups during a simulation. An NVE simulation protocol is developed in which total energy, angular momentum, and linear momentum are conserved. As an application, we present simulation results for the structure and diffusion of a water molecule in hexane.

(1) Praprotnik, M.; Delle Site, L.; Kremer, K. J. Chem. Phys. 2005, 123, 224106. (2) Praprotnik, M.; Delle Site, L.; Kremer, K. Physical Review E 2006, 73, 066701. (3) Praprotnik, M.; Delle Site, L.; Kremer, K. J. Chem. Phys. 2007, 126, 134902. (4) Heyden, A.; Lin, H.; Truhlar, D. G. J. Phys. Chem. B 2007, 111, 2231.

432b Coarse Molecular-Dynamics Analysis Of Structural Transitions In Condensed Matter

Miguel A. Amat, University of Massachusetts, Amherst, 686 North Pleasant Street; 214 Goessmann Laboratory, Amherst, MA 01003, Ioannis G. Kevrekidis, Department of Chemical Engineering, Princeton University, Engineering Quadrangle, Princeton, NJ 08544, and Dimitrios Maroudas, University of Massachusetts, Department of Chemical Engineering, Amherst, MA 01003.

Accurate determination of the onset of structural transitions in complex physical systems is a problem of crucial importance in condensed matter and materials physics. As direct access to such physical responses is typically difficult to attain experimentally, computational techniques such as molecular dynamics (MD) have become powerful tools for probing the underlying atomic-scale dynamics and determining the transition onsets. One of the most attractive features offered by MD lies in its ability to ultimately relate atomistic dynamics to macroscopically observable physical behavior; however, computing the evolution of all of the atomic coordinates over coarse time scales poses a severe limitation to the method. The limitation is compounded when addressing problems with multiple stable states that are separated by high energy barriers. In recent years, novel methods, such as hyperdynamics, transition path sampling, metadynamics, and coarse molecular dynamics (CMD), have been proposed to address long-time dynamics issues directly through atomistic simulation. In CMD, coarse-grained information is estimated on-the-fly from many short and properly initialized independent replica MD simulations. This information can then be used to identify transition points in the physical behavior of the complex systems under consideration. The method is based on the description of the evolution of the probability density, $P(\psi,t)$, approximated by the Fokker-Planck equation where $\psi(t)$ is an appropriate coarse-grained observable that describes the state of the system. In previous studies,[1,2] it has been demonstrated that the CMD method can be applied to determining accurately the thermodynamic melting point of a crystalline silicon model and the critical pressure that marks the onset of a polymorphic structural transition in a crystal under hydrostatic loading.

In this presentation, we report results of CMD analyses of various cases of structural transitions in condensed matter placing emphasis on order-to-disorder and on polymorphic transitions. Specific examples include the melting of bulk metallic crystals and the bcc-to-hcp polymorphic phase transitions undergone by model crystalline materials under high pressure loads. In the melting case, we reconstruct the underlying effective free-energy landscapes over a temperature range and use them to establish a relationship between the melting transition and the loss of structural and elastic stability of the bulk crystal. In the polymorphic transformation application, we extend our previous studies done in the low-temperature limit2 and include effective free-energy landscapes obtained over a wider temperature range. The new results allow determining the limits of inherent stability of the corresponding phases (i. e., bcc and hcp); the outcome of the analysis is a phase diagram, which demarcates the stability boundaries of each phase. We demonstrate that the CMD approach is of general applicability and may be particularly helpful in analyzing various other important types of structural-transition onsets in condensed matter and materials science; selecting appropriate coarse-grained variables is the key ingredient to the successful implementation of this approach through monitoring and analyzing the corresponding, properly initiated coarse-variable evolution.

1. M. A. Amat, I. G. Kevrekidis, and D. Maroudas, Phys. Rev. B 74, 132201 (2006).

2. M. A. Amat, I. G. Kevrekidis, and D. Maroudas, Appl. Phys. Lett. 90, 171910 (2007).

432c Multiscale Simulation Of The Self-Assembly Of Nonionic Poly (Oxyethylene) Alkyl Ether Surfactants

Joshua D. Moore and Keith E. Gubbins. Department of Chemical and Biomolecular Engineering and Center for High Performance Simulation, North Carolina State University, Engineering Building I, Box 7905, 911 Partners Way, Raleigh, NC 27695-7905

The self-assembly of surfactants from a bulk phase onto solid surfaces is of interest in a range of technical and biological processes. New variables associated with the solid surface lead to a phase behavior which is considerably richer than for bulk solutions. Experimentally observed structures include spheres, bilayers, hemi-cylinders and hemi-spheres. Despite the widespread interest in such structures, the underlying principles governing their formation and properties have resisted attempts at theoretical and computational treatments because of the large length and time scales involved. Conventional atomistic computational treatments cannot encompass such a range. However, atomistic interactions such as hydrogen bonding have been shown to be essential to obtain the correct phase behavior. The required length and time scales can be reached using meso-scale methods such as dissipative particle dynamics (DPD), but the atomistic detail is lost. To resolve this dilemma, we employ a multi-scale simulation scheme that allows us to map the fully atomistic system of nonionic poly (oxyethylene) alkyl ether surfactant molecules onto the meso-scale in the bulk and on surfaces. We develop meso-scale effective potentials using directly atomistic trajectories using methods based upon the force-matching approach [1,2]. Comparisons with experiment will be made.

[1] F. Ercolessi and J.B. Adams, "Interatomic Potentials from 1st-Principles Calculations - the Force-Matching Method," Europhys. Lett., 26 583 (1994). [2] S. Izvekov, S. and G.A. Voth, "Multiscale Coarse Graining of Liquid-State Systems," J. Chem. Phys., 123 134105 (2005).

432d Equilibration Of Cystine-Knotted Cyclic Peptides Using Hybrid Monte Carlo

Ramdas S. Pophale[1], Robin Curtis[2], and Michael W. Deem[1]. (1) Bioengineering, Rice University, 6100 Main Street - MS 142, Houston, TX 77005-1892, (2) University of Manchester, PO Box 88, Sackville Street, Manchester, M 60 1QD, United Kingdom

Equilbration of cyclic peptides with disulfide linkages has proved to be difficult. We combine configurational bias Monte Carlo (CBMC) techniques with molecular dynamics and investigate the problem using the hybrid Monte Carlo (HMC). Solvent effects are accounted through solvent accessible surface area and parameter optimized surfaces (POPS) methods. Parallel tempering technique is applied to speed up the equilibration. The peptides of interest are end to end connected cyclic compounds comprising of 30-50 amino acids. Three disulfide bonds forming a cystine-knot is a peculiar feature of the molecules, resulting in slow equilibration. These peptides are known to exhibit specific and potent activity against various pathogens. The hybrid algorithm is expected to facilitate equilibtration significantly better than the existing non-ergodic method.

432e Coarse Molecular-Dynamics Analysis Of Polymorphic Transitions In Crystals Under Uniaxial Loading

Hadrian Djohari[1], Miguel A. Amat[1], Frederick Milstein[2], Ioannis G. Kevrekidis[3], and Dimitrios Maroudas[4]. (1) Department of Chemical Engineering, University of Massachusetts, Amherst, 686 N Pleasant St, Amherst, MA 01375, (2) Departments of Materials and Mechanical Engineering, University of California, Santa Barbara, Engineering II Bldg, Room 2355, Santa Barbara, CA 93106-5070, (3) Department of Chemical Engineering, Princeton University, Engineering Quadrangle, Princeton, NJ 08544, (4) Department of Chemical Engineering, University of Massachusetts, 159 Goessmann Laboratory, 686 N Pleasant Street, Amherst, MA 01003

Determining the onset of elastic instability and analyzing the structural response of crystalline solids beyond the instability onset is a topic of major interest in materials mechanics over a broad class of industrial applications from aerospace to microelectronics. Predicting this limit of strength of crystalline solids requires systematic analyses of the crystals' elastic stability taking all the relevant parameters into account. At any given temperature, the structural response of a crystal to a specified mode of applied mechanical loading will become unstable beyond a critical stress level.

In this presentation, we report results of a coarse-molecular dynamics (CMD) approach to approximating an effective free-energy landscape of the stressed crystalline solids under uniaxial loading and analyzing their stability and stress-induced polymorphic transitions. In CMD, coarse-grained information is estimated on-the-fly from many short and properly initialized independent replica molecular-dynamics (MD) simulations. The method is based on the description of the evolution of the probability density, $P(\psi,t)$, approximated by the Fokker-Planck equation, where $\psi(t)$ is an appropriate coarse-grained observable that describes the state of the system under investigation. The method has been demonstrated to be efficient and accurate in determining the onset of structural transitions in condensed matter, such as melting and pressure-induced polymorphic transitions. In our study, CMD analyses are performed systematically for metallic crystals with a face-centered cubic (fcc) structure at equilibrium subjected to [100] uniaxial loading. The corresponding short MD simulations are based on isothermal-isostress MD according to the Lagrangian formulation of Parrinello and Rahman. Our CMD approach is used to explore a broad range of values of a properly chosen coarse variable. In this analysis, among several options, we have chosen as a coarse variable the stretch along the loading direction, λ_1, which provides a good metric for the deformation of a crystal under uniaxial loading.

By exploring the coarse-variable space, we capture all the possible states of a model metallic crystal, a Morse model for Cu, including fcc, body-centered cubic (bcc), and hexagonal close-packed (hcp) lattice structures. We construct the underlying effective free-energy landscape for every value of the bifurcation parameter (applied uniaxial stress) and construct a comprehensive bifurcation diagram, which includes stable and unstable states, as well as previously unknown branches in comparison with bifurcation diagrams constructed based on lattice-statics analyses. Furthermore, the stability of each branch and the location of "the next stable branch" at the onset of instability also are identified clearly, resulting in the first comprehensive dynamic stability diagram of cubic crystals at finite temperature. Our results also explain why sometimes defects such as stacking faults get introduced into a structure during a transformation mechanism.

432f Simulating Silica Aerogels With A Coarse-Grained Flexible Model And Langevin Dynamics

Lev D. Gelb, Department of Chemistry, Washington University in St. Louis, One Brookings Drive, St. Louis, MO 63130

We present a simple coarse-grained model of silica sol-gel materials which is applied to investigate the formation and structural and mechanical properties of low-density aerogels. This model, unlike many previous developments, is fully flexible and can account for bond breaking and creation upon deformation of the gel structure. We show that direct comparison with experimental structures can be made for appropriate choice of bond-formation control parameters, and that quantitative results for bulk modulus and other properties can be obtained. Additional simulations of syneresis, compression, and extension are performed and microscopically analyzed in order to correlate macroscale behavior with local gel structure and its evolution. The dependence of gel structure and mechanical properties on both "experimental" variables such as reactivity, bond strength, and viscosity, and "simulation" variables such as cell size and integration method, are evaluated. Finally, gas adsorption in fully flexible gels is considered through coupling of a mean-field lattice gas model to the coarse-grained gel model via a perturbation-theory-like approach.

472a First Principle Simulations Of Reactions In Organic Solvents

Veronique Van Speybroeck[1], Michel Waroquier[1], Norbert De Kimpe[2], and Matthias D'hooghe[2]. (1) Center for Molecular Modeling, Ghent University, Proeftuinstraat 86, Ghent, 9000, Belgium, (2) Department of Organic Chemistry, Faculty of Bioscience Engineering, Ghent University, Coupure links 653, Gent, 9000, Belgium

Probably the simulation of reactions in organic solvents is one of the most challenging tasks nowadays in the field of molecular modelling. In many literature works the solvent is just omitted or simulated by using continuum solvation models. In such dielectric model, the solvent is modelled as a continuous medium, usually assumed homogeneous and isotropic, characterized solely by a scalar, static dielectric constant. In some cases, these results give the correct reactive behaviour, provided there are no essential explicit solvent interactions with the solute. In many cases the behaviour of the solvent close to the chemically active part is different from the bulk and specific effects play in the first solvation shell. A possible method to include these explicit solvent interactions might be achieved by placing a few solvent molecules around the chemically active species. Of course then the question arise on how many solvent species to incorporate explicitly. Mostly this number is determined by the value at which the coordination solvation energy is converged. To account also for entropic effects of solvation the Gibbs free energy change of solvation must also be evaluated. [1,2]

A variety of approximations varying from simple gas phase calculation, including solvent implicitly and including solvent with a combined explicit and implicit approach is tested to study the reactive behaviour of a series of aziridinium salts towards regio- and stereospecific ring opening. The aziridine moiety represents one of the most valuable three-membered ring systems in organic chemistry due to its widely recognized versality as a building block towards a large variety of ring opened or ring expanded amines. Unactivated aziridines, i.e. aziridines without an electron-withdrawing group at nitrogen, can easily be transformed into aziridinium salts upon treatment with alkyl halides, which are subsequently opened by a suitable nucleophile (usually the counter ion). As depicted in Scheme 1, the reaction of a chiral aziridine 1 with an alkyl halide may lead to three different amines 4-6 depending on the reaction mechanism. In this way, the initial chiral center at the substituted aziridine carbon atom can be retained (pathway a, amine 4), inverted (pathway b, amine 5) or razemized (pathway c, amine 6). Consequently, the elucidation of the underlying reaction mechanism is of primordial importance when the design of asymmetric syntheses towards chiral amines is contemplated starting from chiral 2-substituted aziridines such as 1.

At first instance the key steps involved in the unprecedented regiospecific ring opening of intermediate 2-(cyanomethyl)aziridinium salts by bromide and pyrrolidine in acetonitrile are investigated theoretically. Experimentally it was found that the ring opening occurred exclusively at the substituted aziridine carbon atom. For the ring opening with pyrrolidine, the correct reaction preference was always found, independent of the solvation model. The quantitative values of the reaction barriers vary substantially with or without solvation. The ring opening with bromide, is found to be more sensitive to the degree of solvation that is taken into account. The correct regioselectivity was only found if a sufficient number of explicit solvent molecules were taken into account and additionally the whole system is placed in a continuum with fixed dielectric constant. Another example which is addressed is the ring opening of 2-(cyanomethyl)aziridines with HBr in acetic acid and acetonitrile or dichloromethane, affording 3-(arylmethyl)amino-4-bromobutyronitriles via regiospecific ring opening at the unsubstituted carbon atom. The factors governing the switch in regioselectivity as compared to the reaction with benzyl bromide are thoroughly investigated. In this case we are dealing with a combination of various solvent species, i.e. acetic acid in combination with acetonitrile or dichloromethane, and thus the initial set up for the cluster model is more complicated. The cluster model is determined on basis of coordination free energies of solvation. The theoretical description of the regioselectivity in the ring opening of C-substituted aziridinium salts is a complex issue, in which the nature of the substrate, the nucleophile and the solvent are be important. These systems constitute ideal examples to test the reliability of a variety of solvent models to describe the correct reaction behaviour.

[1]Matthias D'hooghe, Veronique Van Speybroeck,Michel Waroquier, and Norbert De Kimpe, CHEMICAL COMMUNICATIONS (14): 1554-1556, 2006

[2]Matthias D'hooghe, Veronique Van Speybroeck, Andries Van Nieuwenhove,Michel Waroquier, and Norbert De Kimpe, "Novel synthesis of 3,4-diaminobutanenitriles and 4-amino-2-butenenitriles from 2-(cyanomethyl)aziridines through intermediate aziridinium salts: an experimental and theoretical approach", to be published in J. Org. Chem.

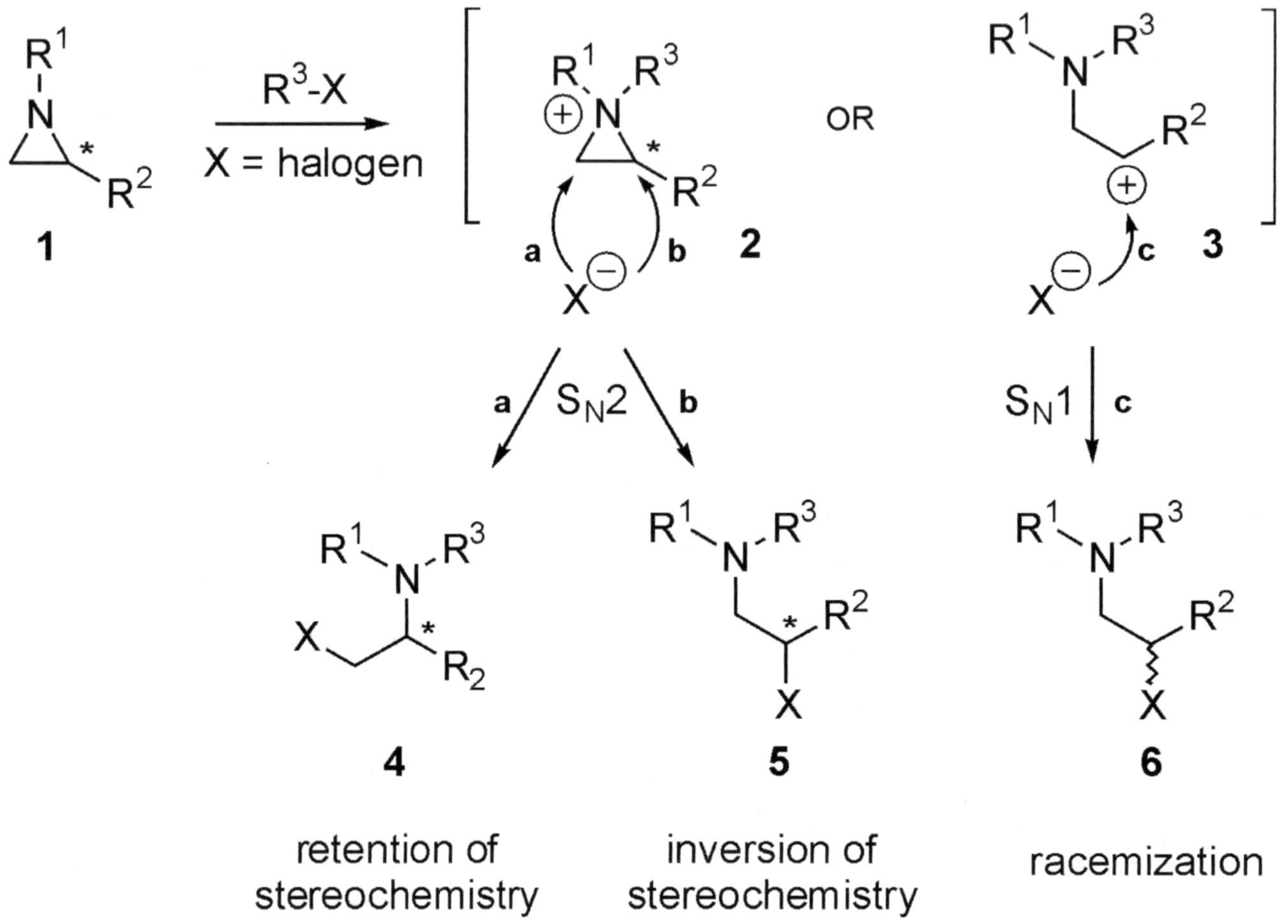

Web Page: molmod.ugent.be

Computational study of surface reaction kinetics and poisoning of metal hydrides

Bing Dai[1], David S. Sholl[2], and J. Karl Johnson[1]

[1] *Department of Chemical and Petroleum Engineering, University of Pittsburgh, Pittsburgh, PA 15261*

[2] *Department of Chemical Engineering, Carnegie Mellon University, Pittsburgh, PA 15213*

The kinetics of reactions is a very important issue one should address when searching for promising hydrogen storage materials. As an example, consider the following family of reactions:

$$2K_3AlH_6 \rightarrow 6KH + 2Al + 3H_2 \quad (1)$$

$$2Na_3AlH_6 \rightarrow 6NaH + 2Al + 3H_2 \quad (2)$$

$$2Li_3AlH_6 \rightarrow 6LiH + 2Al + 3H_2 \quad (3)$$

It is known experimentally that reaction (1) is reversible, reaction (2) can be made reversible with the aid of a catalyst, but experiments have so far failed to observe the reversibility of reaction (3), even with catalysts. It is not known what is responsible for the differences in this family of reactions. One possible reason for the differences in reversibility of the reactions could be differences in the surface reactivities of the respective hydrides. For example, it may be that the LiH surface is much more susceptible to poisoning by O_2 or H_2O than the other two surfaces and that may account for the differences in reversibility. Therefore, we have used first principles density functional theory to study the adsorption and dissociation of H_2, O_2 and H_2O gas phase molecules on the surfaces of alkali metal hydrides (LiH, NaH and KH). This is regarded as a first step in understanding the reaction kinetics of this family of alkali metal alanates.

Adsorption of atomic hydrogen is energetically unfavorable on all three of the metal hydride surfaces, when compared to hydrogen in the gas phase. This indicates that hydrogen should not react with the surface at low H_2 partial pressures. We have also computed the adsorption energies for molecular hydrogen on the MH surfaces. Our calculations show that molecular hydrogen weakly physisorbs on each of the surfaces, with binding energies of a few kJ/mol.

We have computed binding energies and geometries for adsorption of atomic and molecular oxygen on the three metal hydrides surfaces. Both atomic and molecular oxygen adsorb on the surfaces with energies in the chemisorption range. Optimization of atomic oxygen on LiH at 1 monolayer coverage results in a surface oxidized structure that strongly resembles the structure of LiOH. Interestingly, we found that O_2 molecules dissociates without a barrier on LiH and NaH surfaces, as long as the coverage of O_2 is low. Molecular oxygen dissociates to form OH groups with the surface H atoms. In contrast, we did not observe the barrierless dissociation of O_2 on the surface of KH.

We have found that H_2O is weakly chemically bound to the LiH, NaH, and KH surfaces. The H_2O dissociation pathways on these surfaces have also been calculated. The barrier heights between 14 to 25 kJ/mol for all three systems, after zero point energy corrections. The relatively low reaction barriers indicate that H_2O will dissociate very rapidly at room temperature on all the three surfaces.

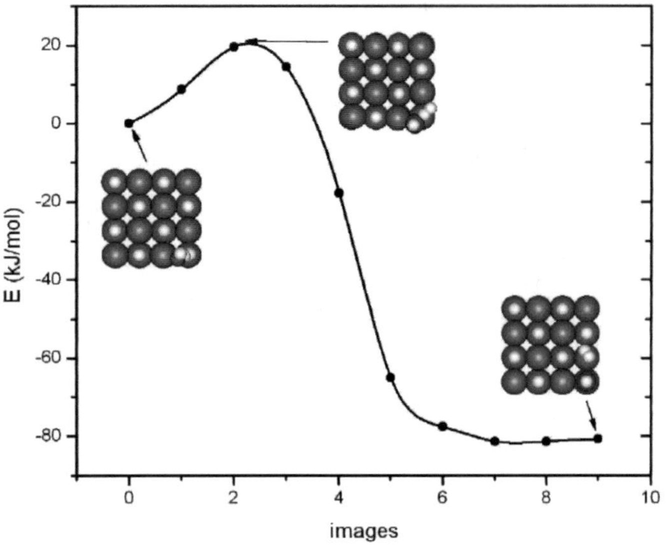

Figure 1: The energy dissociation pathway for H_2O on the NaH surface.

The paper is only the first step toward the understanding of the kinetics of solid phase hydrogenation of complex hydrides. We can conclude that the differences in reversibility of the reaction (1), (2) and (3) are not likely due to differences in the surface reactivities of the respective hydrides. Further kinetic studies of these systems are clearly needed before the differences in reaction reversibilities can be resolved.

472c Thermodynamics From First Principles: Phase Diagramm And Thermodynamic Properties Of Oxygen Adsorbed On Ni(111)

Frerich J. Keil, Department of Chemical Engineering, Hamburg University of Technology, Hamburg, D-21073, Germany and Cesar Lazo, Chemical Engineering, Hamburg University of Technology, Hamburg, D-21073, Germany.

Abstract The interaction of oxygen (O) with metal surfaces is the basis for a number of important technological processes such as bulk oxidation, corrosion, and heterogeneous catalysis, and has thus been studied in great detail, both from a macroscopic and a microscopic point of view. In the present study we calculated the O/Ni(111) phase diagram and thermodynamic properties, like internal energy, specific heat, free energy, and entropy, based on first principles. Lateral interaction parameters were obtained from density functional theory (DFT, full potential LAPW/APW+lo)) calculations employing techniques from statistical mechanics, i.e. the cluster expansion method. The statistical mechanics problem was numerically solved using Metropolis Monte Carlo simulations in the canonical and grand canonical ensemble, using Kawasaki and Glauber sampling, respectively. To measure continuous order-disorder phase transitions in canonical MC simulations, order parameters are defined according to the symmetry broken during the transition. In order to locate first-order phase transitions, simulations in the grand canonical ensemble were carried out. Scans were performed at constant temperature and chemical potential. For first-order phase transitions simulations in the canonical ensemble can lead to wrong results due to finite size effects and phase coexistence. Additionally, the Wang-Landau Monte Carlo simulations in the canonical ensemble were used to study the nature of the phase transitions at various surface coverages. By shifting the acceptance rule from energy to entropy space, the Wang-Landau algorithm is able to surmount the problem posed by hysteresis at first-order phase transitions and critical slowing-down at continuous phase transitions.

Careful investigations were executed on the proper statistics for obtaining the interaction energy parameters of the lattice-gas Hamiltonian taken from plane-wave DFT calculations.

The experimental phase diagram and thermodynamic properties could quite well be simulated. All phases that have been experimentally determined are present in our simulated phase diagram. Additionally, we found a coexistence region of p(2x2) and (√3x(√3)R30° phases at very low temperatures. Limitations in the accuracy of the calculated temperatures will be discussed.

472d Quantum Dots On Al(110): New Lattice Models For Thin-Film Growth Based On First Principles

Yogesh Tiwary, Dept. of Chemical Engineering, The Pennsylvania State University, 174 Fenske Laboratory, University Park, PA 16802 and Kristen A. Fichthorn, Dept. of Chemical Engineering, The Pennsylvania State University, 164 Fenske Laboratory, University Park, PA 16802.

Atomic self-organization at the nanometer scale is a complex interplay between kinetic and thermodynamic phenomena. This interplay is evident in Al(110) homoepitaxy. In experimental studies of Al(110) homoepitaxy, it is observed that over a certain temperature window (330-500K), three-dimensional nanohuts form and self-organize over the micron scale [1,2]. The factors leading to this self-organization are currently unclear.

Using ab initio calculations based on density-functional theory with VASP, we determined the pair and multi-body interactions between Al adatoms on this surface. Within the framework of the lattice-gas model, we find that multi-body interactions (four-body, five-body, and higher) are significant. However, lattice-gas models become complicated with such high-order multi-body interactions. We developed a simple and accurate model to estimate the energies of clusters of any size. Employing this model, we see a transition from linear chains to rectangular islands as we increase the cluster size.

We used the climbing-image nudged elastic band method [3] to calculate the energy barriers for surface diffusion on Al(110). Exchange diffusion is seen to be the favorable mechanism. In the presence of neighboring adatoms, the activation energy changes due modifications of the initial, transition, and final state energies. In previous work, simple models have been used to estimate the effect of neighboring adatoms from the activation energy for isolated adatoms and the energy of the initial state [4] or from both the initial- and final-state energies [5]. These simple models cannot accurately describe the effect of neighboring adatoms on exchange. We present a model that considers all possible paths for exchange and provides the energy barrier by taking into account the actual transition-state configurations. Results from kinetic Monte Carlo simulations show that our model predicts a higher island density than previous ones, in agreement with experiment. The insight gained in this study could be employed to accurately predict nano-scale self-assembly in different growth scenarios.

[1] F. Buatier de Mongeot, W. Zhu, A. Molle, R. Buzio, C. Boragno, U. Valbusa, E. Wang, and Z. Zhang, Phys. Rev. Lett. 91, 016102 (2003).

[2] K. Fichthorn and M. Scheffler, Nature 429, 617 (2004).

[3] G. Henkelman, B.P. Uberuaga, and H. Jonsson, J. Chem. Phys. 113, 9901 (2000).

[4] A. Videcoq, F. Hontinfinde, R. Ferrando, Surf. Sci. 515, 575-587 (2002).

[5] K. A. Fichthorn and M. Scheffler, Phys. Rev. Lett. 84, 5371 (2000).

472e Structure, Interconversion And Reactivity Of Point-Like Surface Defects On Amorphous Silica

Chin-Lung Kuo, Min-Chan Kim, and Gyeong S. Hwang. Chemical Engineering, The University of Texas at Austin, 1 University Station Stop C0400, Austin, TX 78712

Atomic-level control and accurate determination of the structure and properties of oxide surfaces has long been an issue of great importance due to their many applications in electronics, heterogeneous catalysis, and electrochemistry. However, the difficulty of direct characterization arising from sample charging has impeded progress towards understanding the fundamental behavior and properties of oxide surfaces. Point-like defects may significantly alter the surface properties of amorphous silica, but their formation, structure, and dynamics are still unclear. Earlier experimental investigations have demonstrated that upon electron irradiation a silica surface undergoes changes in not only photosensitvity but also surface reactivity, mainly due to irradiation-induced defects. Oxygen deficient centers (ODCs) are most common point defects in oxide materials While a great deal of effort has been devoted to understanding the nature of ODCs in bulk silica, little is known about the formation, structure, and dynamics of surface defects despite their importance in determining complex silica surface chemistry. In this talk, we present the structure and stability of oxygen vacancy related defects on amorphous silica based on combined classical Monte Carlo and first principles quantum mechanical calculations. Based on model amorphous silica surfaces constructed using combined Monte Carlo and density functional calculations, we determined the structure of point-like surface defects, such as Si-Si dimer, divalent Si, and silanone, and the pathways and barrier their interconversion using extensive ab initio molecular dynamics and static transition-state search calculations. These results suggest that surface oxygen vacancies in the dimer configuration may undergo thermally-activated transformation with moderate activation energies into other stable point-like defects including divalent Si defects, silanone groups, and/or subsurface Si-Si dimers. This in turn implies that under thermal equilibrium the surface concentration of oxygen vacancies in the dimer configuration would be minimal, while providing a theoretical support for the existence of divalent Si defects and silanone groups on the amorphous surface. We will also discuss the reactivity of these surface defects towards various gas species. The detailed understanding greatly assists in understanding the defect-associated complex surface chemistry of amorphous silica.

Using First-Principles Calculations and Cluster Expansions to Model Hydrogen Diffusion in Metal Alloys

Lymarie Semidey-Flecha and **David S. Sholl**, Carnegie Mellon University, Department of Chemical Engineering, Carnegie Mellon University, Pittsburgh, PA 15213

Diffusion of hydrogen in metals is important in numerous current and future applications of hydrogen, including hydrogen purification and storage. While plane wave Density Functional Theory calculations are well suited to giving local site-to-site information on diffusion paths and rates, this information is insufficient to describe net diffusive transport in disordered alloys. We will discuss how first-principles calculations, cluster expansion methods, and Kinetic Monte Carlo simulations can be combined to rigorously describe the net diffusion of interstitial in disordered alloys. Success in this task requires efficient use of first-principles methods, since hundreds or thousands of individual site energies and hopping rates are needed. We will discuss methods that allow us to rapidly locate hundreds of distinct transition states for diffusion of interstitial H in metals.

472g First-Principles Theoretical Analysis Of Carbon Nanostructures And New Carbon Allotropes

Tejinder Singh[1], Michael J. Behr[2], Mayur S. Valipa[1], Eray S. Aydil[2], and Dimitrios Maroudas[1]. (1) University of Massachusetts, Department of Chemical Engineering, Amherst, MA 01003, (2) Department of Chemical Engineering and Materials Science, University of Minnesota, Minneapolis, MN 55455

Diamond is a technologically important material due to its unique properties such as extreme hardness, high electrical resistivity, high thermal conductivity, and chemical inertness. It is used widely in the fabrication of machine tools, optical coatings, X-ray windows, and light-emitting optoelectronic devices. The synthesis of diamond or diamond-like thin films at room temperature by chemically-driven allotropic transformations has evinced keen interest recently; one such method to synthesize such films involves the treatment of carbon nanotubes (CNTs) and carbon nanofibers (CNFs) with H atoms from an H_2 plasma. However, the development of systematic strategies for depositing thin films of diamond with desirable properties requires a fundamental understanding of the interactions between H atoms and the growth surface, which remains unclear. In this context, we aim at analyzing the crystalline C phases that are observed upon exposure of CNTs to an H_2 plasma and to investigate the role of H in determining the structure of the transformed C films.

This presentation focuses on a detailed analysis of the various crystalline phases of C observed upon exposing CNTs to an H_2 plasma, as well as the potential role of H in determining the structures of the resulting C nanocrystals. In our studies, we have combined experimental characterization tools based on high-resolution transmission electron microscopy (HRTEM) and electron diffraction with first-principles density functional theory (DFT) calculations. Our DFT calculations are carried out within the generalized gradient approximation and employ plane-wave basis sets, ultra-soft pseudopotentials, and slab supercells. Our experiments reveal that, upon exposure of CNTs/CNFs to an H_2 plasma, a transformation is induced of the graphitic crystalline network to an amorphous carbon matrix, along with the formation of a web of crystallites (~ 5-10 nm in diameter) that are found embedded in the amorphous matrix. Subsequent HRTEM and electron diffraction studies have revealed the presence of two distinct crystalline phases embedded within the amorphous matrix. The observed electron diffractions and lattice spacing from these two phases are consistent with a face-centered cubic (fcc) lattice with lattice parameter, a, of 4.25 Å and a body-centered cubic (bcc) lattice with a = 3.0 Å; these structures are distinct from pure fcc carbon and bcc carbon phases, respectively.

Using DFT calculations, we have analyzed the structure of several new allotropes of pure carbon, which are candidate structures to explain the experimentally observed carbon phases. Our results indicate that, in the absence of H in the crystal structure, none of the allotropes can account for the lattice parameter of the observed carbon phases. Hence, we consider the incorporation of H in the interstitial sites of the cubic phases of carbon, and investigate the effect of varying the composition, C_xH_y (x,y < 3), of the carbon phases. Our calculations show that the bond-center (BC) configuration is the most stable interstitial H defect configuration in diamond carbon; in fact, the carbon lattice expands in order to accommodate the extra H atoms located at BC sites. However, there is a critical H concentration within the diamond lattice that onsets the stability of the C_xH_y lattice structure. From a comparison of the formation energies of the C_2H and CH structures, we infer that there exists an intermediate configuration, with a lattice parameter of ~ 4.27 Å, which renders the C_xH_y lattice structure stable; this stable structure also can provide an explanation for the experimentally observed fcc phase. Similar analysis performed for the bcc phase reveals that there exists an intermediate configuration, with composition that varies between that of the CH (a ~ 2.90 Å) and CH_2 structures (a ~ 3.10 Å), whose lattice parameter can account for that of the observed bcc phase. Hence, based on our theoretical analysis of C structures and allotropes, we conclude that the experimentally observed carbon nanostructures can be explained by accounting for the presence and concentration of H in the C lattice.

561a The Problem And The Benchmark Data

James Olson, Analytical Chemistry, The Diow Chemical Company, 770 Building, South Charleston, WV 25303

A description of the simulation challenge problem and the benchmark data used to judge it.

561b Announcement Of The Champion And The Next Challenge

Richard B. Ross, Corporate Research Materials Laboratory, 3M Company, 201-2E-23, 3M Center, St. Paul, MN 55144

After presentations from the entrants, the Champions will be announced. Also, the next simulation challenge problem will be announced.

592a Molecular Simulations of Polycyclic Aromatic Compounds

Carlos Nieto-Draghi, Benoit Creton, Antony Bocahut, Theodorus De Bruin, and Veronique Lachet. Thermodynamic and Molecular Simulation Department, IFP, 1 et 4 Avenue de Bois-Preau, Rueil-Malmaison, 92852, France

Abstract

Detailed information of thermodynamic properties like density, vapour pressure and vaporisation enthalpy, and dynamic properties as viscosity, of fuels in diesel engines under working conditions are hard to obtain, as a result of the high temperatures (up to 750 K) and pressures (250 MPa) that reign in the engine, making experimental measurements almost impossible to realise. Recently, it has been shown for the benzene molecule that empiric simulations applying Anisotropic United Atom (AUA) intermolecular potential need to explicitly take into account electrostatic interactions to obtain accurate viscosities [1].

Extension of this approach to polycyclic aromatic hydrocarbon molecules (with cycles ranging from 2 to 4, cf. Figure 1), which are important constituents of diesel fuels is now under progress. Electrostatic interactions were included using point charges that were derived from renormalized quadrupole moments based on quantum mechanical calculations.

Figure 1

In this communication we report the systematic optimisation of these point charges in order to reproduce thermodynamic and dynamic properties of the pure compounds, predicted by means of Gibbs ensemble Monte Carlo and MD simulations and compared to the experimental data available. Finally, we have used the optimised potential to predict the viscosity of polycyclic aromatic molecules for which no experimental data are available.

[1] Bonnaud P.; Nieto-Draghi C.; Ungerer P. *J. Phys. Chem. B* **2007**, 111, 3730.

592b Understanding the Relationship between Molecular Architecture and Thermophysical and Rheological Properties of Perfluoropolyethers

Bangwu Jiang[1], Jamie L. Adcock[2], Nathan J. Crawford[3], Alviony F. Pangloli[3], Jared Fern[3], Brian J. Edwards[3], William V. Steele[3], and David J. Keffer[3]. (1) Chemical Engineering, The University of Tennessee, 419 Dougherty Hall, Knoxville, TN 37996, (2) Department of Chemistry, The University of Tennessee, Knoxville, TN 37996, (3) The University of Tennessee, 419 Dougherty Hall, Knoxville, TN 37996

Understanding the relationship between molecular architecture and thermophysical and rheological properties of perfluoropolyethers

Abstract Perfluoropolyethers (PFPEs) form a class of lubricants, which are broadly applied in oxygen service, in aircraft instrument bearings, in reactive chemical environments, in vacuum pumps, in sealed-for-life electric motors, in computer hard drives, and as high-temperature greases. The relationship between molecular architecture and physical properties of PFPEs is crucial to design the high-performance PFPE lubricants with revised structure. In this work, the following five short-chain PFPEs with different structures have been synthesized. The chains vary four key architectural features including (1) monomer size (-CF_2O-, -CF_2CF_2O-, -$CF_2CF_2CF_2O$- , (2) chain length, and (3) end-group size.

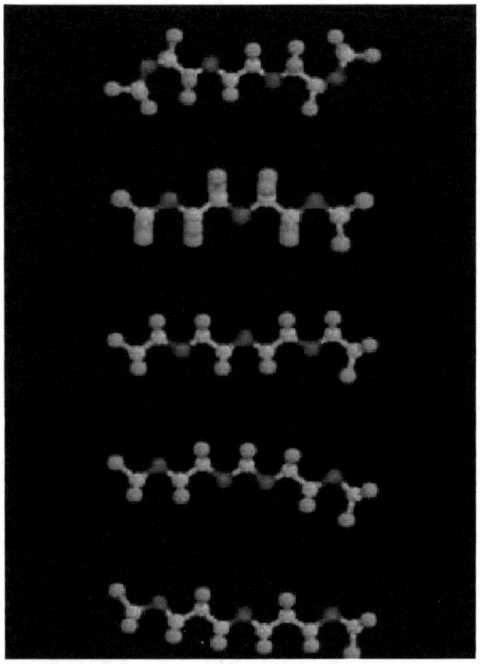

Snapshot of five short-chain PFPE molecules

Previous work focused on the first four PFPEs and led to a revision of the Universal Force Field (UFF)[1] Model to allow for different types of fluorine atoms, depending upon their local environment, which resulted in a quantitative description of the relationship between the molecular architecture of the lubricant and the viscosity [2].

In this contribution, we extend the work in two ways. First, we have performed Equilibrium Molecular Dynamics (EMD) and Nonequilibrium Molecular Dynamics (NEMD) simulations to test the predictive capabilities of the new potential. The additional compounds include (i) compounds with variation in monomer type and (ii) compounds of greater chain length. For the longer chain PFPEs, we study the effect of chain length on viscosity in these systems, and how a longer chain deforms under a strong flow field. Bulk rheological properties of the perfluoropolyethers are investigated as functions of temperature, shear rate, and molecular structure. Structural properties such as the mean-square end-to-end chain lengths and radii of gyration are collected and analyzed as functions of shear rate and molecular architecture. Rheological properties from molecular simulation are compared to the experimental data.

Second, experimentally determined PVT data, phase diagrams, heat capacities, critical properties, and viscosities are reported. Again, we examine the relationship between these thermodynamic properties and the molecular-level structure observed in the EMD simulations.

Acknowledgements:

This work is supported by Air Force Office of Scientific Research through contract # FA 9550-05-1-0342. The authors wish to acknowledge resources of the Center for Computational Sciences at Oak Ridge National Laboratory, which is supported by the Office of Science of the DOE under Contract DE-AC05-00OR22725.

(1) Rappe, A. K.; Casevit, C. J.; Colwell, K. S.; Goddard, W. A.; Skiff, W. M. *J. Am. Chem. Soc.* **1992**, *114*, 10024.

(2) Jiang, B., Crawford, N.J., Keffer, D.J., Edwards, B.J., Adcock, J.L., *Mol. Simul.* **2007**, *accepted* .

592c Modeling Permeability Of Water And Chemical Agents In Sulfonated Triblock Copolymers

Jan Andzelm, James Sloan, Eugene Napadensky, Dawn Crawford, and Margaret Hurley. US Army Research Laboratory, Aberdeen, MD 21005

The sulfonated polystyrene-b-polyisobutylene-b-polystyrene copolymer (sSIBS) is of particular interest because of potential applications for fuel cell and textile applications, where breathable yet protective clothing is required. In this work, we have used multiscale modeling to gain understanding of static and dynamic properties of this polymer at detailed atomistic and mesoscale level. We have used a polymer model of a highly sulfonated material at 64% of sulfonation neutralized with monovalent (Na,Cs) divalent (Mg,Ca, Ba) and trivalent ions (Al). The dynamics of cross linking due to the metal present and the permeability of water were studied. The diffusion coefficients of water were calculated and compared with the recent experimental data. It was found that the presence of metal ions lowers the diffusion coefficients. The mechanism of diffusion and effect of membrane architecture will be discussed. Metal ions form rather stable 2- and 3- cross links with neighboring sulfonated groups. To probe the protective capabilities of this polymer, simulations were performed of diffusion of dimethylmethylphosphonate (DMMP) and bis (2-chloroethyl)sulfide (sulfur mustard) within the polymer framework. The Steered Molecular Dynamics (SMD) capability of the NAMD software was utilized to probe events on timeframes not accessible through standard dynamics simulations. The effect of the metal ion on DMMP and sulfur mustard transport will be discussed.

592d Molecule Relaxation And Viscosity In Model Asphalt Mixture Systems

Liqun Zhang and **Michael L. Greenfield**. University of Rhode Island, Dept. of Chemical Engineering, 205 Crawford Hall, Kingston, RI 02881

Mixtures resembling asphalts were studied using equilibrium molecular dynamics in order to understand the contributions of individual molecules to overall viscosity. Model asphalts were devised based on solubility and polarity, using one or more compounds to represent each of the four typical classifications of molecules found in real asphalts (asphaltene, polar aromatic, naphthene aromatic, saturate). Relaxation dynamics were assembled using rotational correlation functions of aromatic ring normal vectors and end-to-end vectors. Typical relaxation behavior was estimated using modified KWW fits of the simulation results. A Vogel-Fulcher temperature dependence of the relaxation time was observed, and mixture viscosity over a range of temperatures was then estimated using a high-temperature direct calculation (via the Green-Kubo method) and the Debye-Stokes-Eistein relationship. The same methods were applied to both unmodified and polymer modified model asphalt mixtures. Viscosity results at room temperature agreed reasonably with literature values. The mechanism of how polymer chains modify asphalt viscosity was also analyzed. The long-term results can supply guidance for desiging new functional polymers for use in road pavements.

592e Enthalpy/entropy Effects In The Self-Assembly Of Nanoparticle Complex Systems: A Multiscale Molecular Modeling (M3) Approach

Maurizio Fermeglia and Sabrina Pricl. Molecular Simulation Engineering Laboratory, Department of Chemical Engineering, University of Trieste, Piazzale Europa 1, Trieste, 34127, Italy

Nanocomposite materials are receiving increased attention for potential applications including magnetic storage media, high-surface-area catalysts, selective membranes, and photonic band gap materials, etc. Their role in more conventional applications is still relevant in many different fields. In fact, this combination of classical organic polymers or hybrid organic/inorganic matrices and nanoparticles can lead to the enhancement of mechanical, optical, thermal, fire-retardant, and ablative properties, as well as gas-transport properties compared to either of the individual components). In particular, due to the ability to microphase separating into a variety of ordered structures on nanometric length scales, these matrices can be used as ordered template to afford opportunities for controlling the spatial and orientational distribution of nanoparticles to lead to new materials with tailored microstructure-dependent properties. Whatever the matrix, it is of paramount importance to describe the fundamental phenomena that leads to the microstructure at nanoscale level and, for achieving this result, it is necessary to resort to multiscale molecular modeling. The microphase behavior of complex nanoparticle systems may be described by mesoscale simulation methods. Among these, a particle based (DPD) method is chosen here for the description of the system since it considers beads of different types and different dimensions that allow us to simulate both the enthalpic and entropic effect of the nanoparticles embedded in a polymeric matrix. Having obtained a realistic morphology at mesoscale, it is possible to transfer the simulated mesoscopic structure to finite elements modeling tools (FEM) for calculating macroscopic properties for the systems of interest. If all these methods are tightly integrated, a complete, multiscale molecular modeling (M3) recipe for studying and characterizing these materials is obtained. In this paper we apply this procedure to systems made up by different polymeric and hybrid organic/inorganic matrices filled with nanoparticles of different chemical nature, size, and shape, thus accounting for both enthalpic and entropic factors. We will present an integrated multiscale method that starts at atomistic level with the determination of the interactions energies between the components involved, passes this information to the mesoscale level where the morphology is simulated including the embedded particles, and finally the macroscopic properties are calculated with a finite elements method. Simulated data at each single scale will be compared with experimental information to validate the methodology.

Web Page: www.mose.units.it

Construction and Study of Porous Polymer Media for Ion-Exchange Adsorption by Molecular Dynamics

E. Riccardi, J.-C. Wang, and A. I. Liapis, University of Missouri-Rolla, Rolla, MO 65409-1230

Ion-exchange chromatography (IEC) employing porous adsorbent particles in which the charged affinity group/ligand (adsorption site) is linked to the base matrix via a polymeric extender, plays a significant role in downstream purification of produced biomacromolecules. The properties of the discrete porous structure formed by the polymeric extender attached on the base matrix are of crucial importance because they can affect (i) the selection of the charged affinity group/ligand that is to be immobilized on the surface of the of the pores of the porous polymeric structure, (ii) the transport of the selected charged affinity group/ligand in the pores of the polymeric layer; (iii) the immobilization and distribution of the adsorbent sites on the surface of the pores of the porous polymeric structure, and (iv) the transport, rate of adsorption, and adsorption loading of the adsorbate of interest. However, the behavior of the porous structure of the adsorbent particles and its effects on the transport and adsorption of biomacromolecules cannot be fully determined by independent experimental measurements or macroscale models. In this regard, molecular dynamics (MD) modeling and simulation studies have the potential to provide useful scientific information and knowledge to significantly expand our understanding and improve our ability to develop better porous adsorption media with high adsorption rates and capacities for IEC separation of desired biomolecules. In this work we aim to develop tractable model systems for MD simulations that represent polymeric extender layers having desirable pore sizes to accommodate charged affinity groups/ligands and charged adsorbate molecules of interest. Such systems can also serve as base systems for future MD studies to characterize the effects of charged affinity groups/ligands on the properties of pore structure and investigate biomolecular transport in the pores of the porous layers and adsorption onto the adsorbent sites.

Suggested by experimental findings and for computational efficiency, our simulation system consists of 10 dextran chains immersed in water and immobilized on an agarose surface. In the lateral directions, the linear dimensions of the simulation system are 100Å×100Å and periodic boundary conditions are applied. Each of the dextran chains has 40 monomers and 5% branching with 1 or 2 side units. We modified the M3B model, a rigorous coarse grain model developed for malto-oligosaccharides, to simulate dextran chains. To represent the aqueous phase, 32000 water molecules are included and simulated as coarse grain beads with proper size and interaction strength. The simulated surface is effectively a curved one with a flat base surface treated by the Steele's 10-4-3 model and a few large spheres embedded at the topmost layer. Dextran chains are anchored on the spheres at randomly selected locations by immobilizing their first monomers and with their axes perpendicular to the sphere surface. This novel use of spheres, or more generally similarly curved surfaces, gives the simulation studies several important advantages. The most important one is the introduction of an additional length scale or control variable into the simulation system that could generate proper pores of pre-determined sizes to accommodate a range of biomolecules without significant increases in the linear dimensions and number of simulated water molecules.

In our simulations, the dextran chains started with linear conformations and evolved into helical structures as shown below. The consistency of the helical structures with the conclusions of existing studies indicates the adequacy of our dihedral/torsional potential models. Through mutual interactions and interactions with the surface, the dextran chains collapse against each other and form pore structures that are stable within the time frame of our simulations.

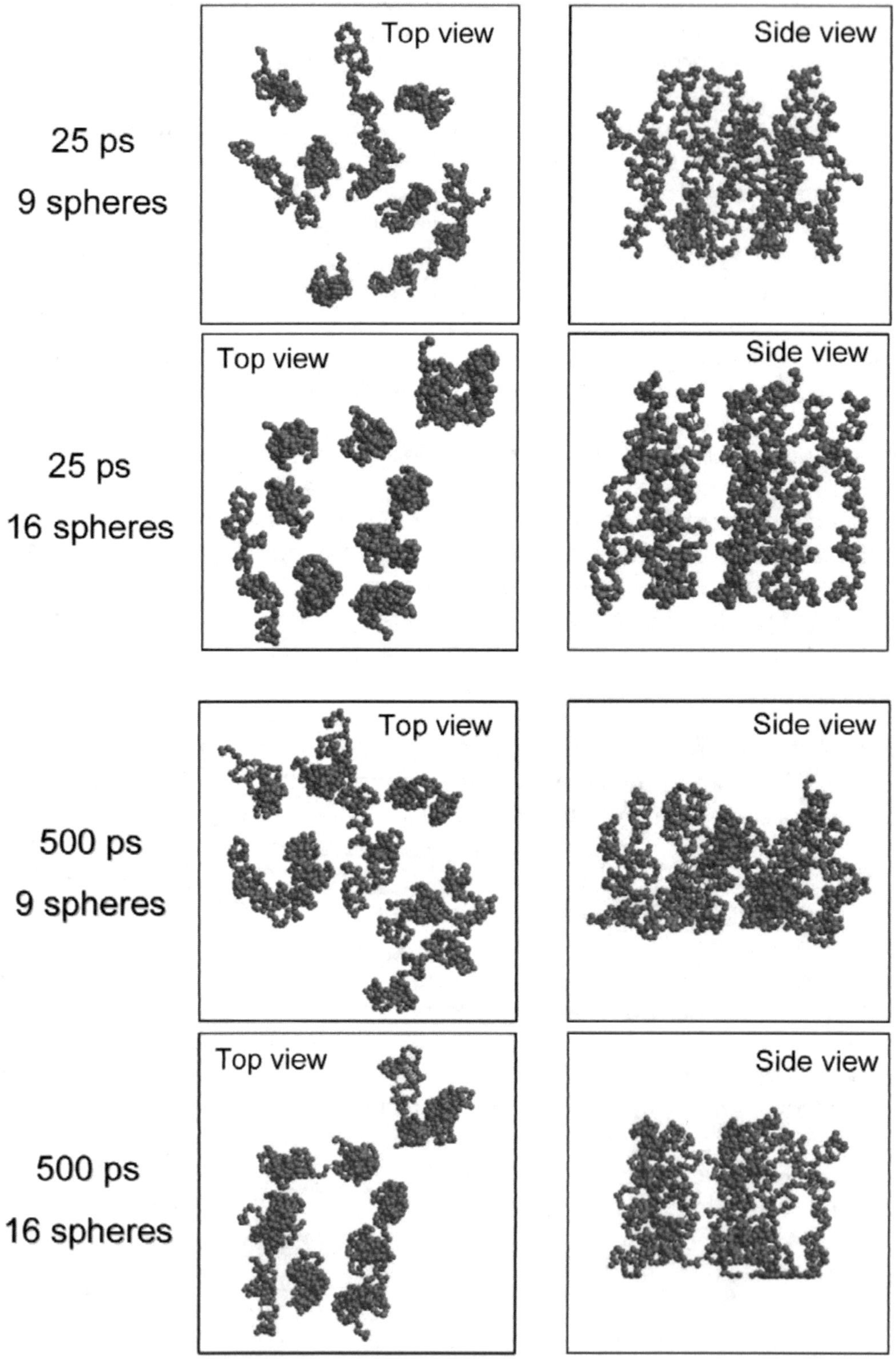

To characterize the resultant pore structures, we divide the simulation systems into a lattice of cubes whose dimension along each of the x, y, and z directions is 3.3 Å, which is approximately equal to the size of a water molecule. Each cube is then further divided into a lattice of cubic cells whose dimension is 0.3 Å. By knowing the positions of the beads of the dextran polymer chains, spherical probes of varying diameters are employed to navigate through the two-level cubic lattices to evaluate the opening and volume of the pore space. The emphasis is on the largest pores that are suitable for biomolecules. The size distributions of larger pores of the two simulation systems studied are tabulated below, where it should be mentioned that larger pores are located closer to the outer surface of the dextran layers. It is apparent from the comparison that within the same linear dimensions, using fewer embedded spheres that have larger diameters result in more pores greater than 20 Å.

In conclusion, it was demonstrated that a combination of flat surface and embedded spheres can be used to construct tractable porous polymer model systems with desirable pore structures in order to accommodate biomolecules of interest without having to simulate systems with significantly larger linear lateral dimensions. The simulation results confirm that larger in diameter spheres create close to the outer surface of the polymer layer more larger pores than those generated when smaller in diameter spheres are employed. The simulation systems can also be sued as base systems to attach ligands and adsorb biomolecules. These system steps are denoted as Stage II and Stage III, respectively, in our research project.

Pore Radius (Å)	Pore Frequency (9 spheres)	Pore Frequency (16 spheres)
14 - 15	2	3
15 - 16	1	1
16 - 17	1	1
17 – 18	1	2
18 - 19	0	1
19 - 20	1	2
20 - 21	0	0
21 - 22	1	0
22 - 23	1	1
23 - 24	1	0
24 - 25	0	0
25 - 26	0	0
26 - 27	0	0

592g Using Elementary Reactions To Model Growth Processes Of Polyaromatic Hydrocarbons Under Pyrolysis Conditions Of Light Feedstocks

Karen Hemelsoet[1], Veronique Van Speybroeck[1], Kevin M. Van Geem[2], Guy B. Marin[2], and Michel Waroquier[1]. (1) Center for molecular modeling, Ghent University, Proeftuinstraat 86, Gent, 9000, Belgium, (2) Laboratorium voor Petrochemische Techniek, Ghent University, Krijgslaan 281, S5, Gent, 9000, Belgium

Steam cracking or pyrolysis of hydrocarbons remains one of the simplest and oldest petroleum refinery processes and is considered as the main production source of light olefins such as ethylene and propylene. These species form key products in modern chemistry. Simultaneously with the cracking processes, the formation of a carbonaceous residue (or coke) on the inner walls of the reactor tubes has been observed. The presence of this carbon-rich structure is highly undesirable as the coke layer leads to a higher pressure drop over the reactor and an increased heat-transfer resistance. The coke growth mechanism consists of different stages: initially coke is formed by a heterogeneous catalytic mechanism involving the metal tube skin of the cracking reactor. Once the surface is covered with a thin filamentous layer a heterogeneous non-catalytic process takes over, resulting in a graphite-like structure. In particular circumstances (at very high temperatures or using heavy feedstocks) a homogeneous, non-catalytic mechanism is also observed. It is reported that for the temperature intervals relevant in the steam cracking units, the second mechanism is the most important. Due to its graphite-like characteristics, the coke layer can be modeled as a network consisting of aromatic rings with varying carbon/hydrogen ratio. The overall cracking and coke formation processes are known to proceed through a complex cascade of free radical reactions. Due to this complexity, simulation models have become essential tools in today's operation plant. Recently, the effect of radial gradients (for the temperature as well as molecular and radical concentrations) on the global coking rate was investigated using a 2-dimensional reactor model [1] and the correspondence with experimental data was shown to be satisfying.

In order to study the growth of the coke layer, a model based on elementary reactions was proposed [2]. The main advantage of such a model is the fundamental nature of the elementary steps and consequently its general applicability. In the particular model, five classes of reversible reactions can be distinguished: hydrogen abstraction; substitution; gas phase olefin addition to radical surface species; gas phase radical addition to olefinic bonds; and cyclization. The complexity of the mechanism prohibits a complete experimental investigation and therefore, ab initio modeling techniques are applied to provide reliable kinetic data about important reaction steps. Several of the distinct reactions have already been examined using advanced ab initio methods (an overview is given in ref. [3]). In particular, the recently developed hybrid BMK functional as well as composite methods (using post-Hartree-Fock energies) have been proven successful for the important class of hydrogen abstraction reactions [4,5]. The former method is generally preferred for systems of (moderately) large size since density functional theory (DFT) provides an excellent cost-to-performance ratio.

To identify the elementary reaction classes that determine the global coking rate, microscopic routes that start from benzene and lead to naphthalene have been investigated. It is found that initial hydrogen abstraction reactions (with or without assisting radicals) and final dehydrogenation reactions are decisive. Overall, the concentrations of the involved species are found to be crucial to verify the most possible reaction pathway. In this light, the low concentration of available gas phase radicals (e.g. methyl radical) strongly influences the observed low rate constants of the bimolecular abstraction reactions. In this work, two new routes are proposed and thoroughly investigated using high-level DFT-methods. One route starts from the benzene molecule, whereas in the other route the focus lies on toluene and its derivatives. Intrinsic as well as modified reaction rate constants for the elementary steps in both schemes are presented.

The influence of the local polyaromatic structure on the kinetics of the hydrogen abstraction reactions is studied by performing calculations on larger polycyclic structures. Hydrogen abstractions at aromatic structures involving six-membered rings and leading to aryl and benzyl radicals (in the case of the benzene, toluene route, respectively) have been examined. In the first case, this has revealed six types of possible reactive site at the coke surface, with the local polyaromatic structure strongly influencing the reaction kinetics [5]. Overall, we find that hydrogen abstraction preferentially occurs from the surface sites of smaller PAHs, such as benzene and naphthalene.

References

[1] Van Geem K. M. , Heynderickx G. J. and Marin G. B. , AIChE J. 2004, 50, 173. [2] Wauters S. and Marin G. B. , Ind. Eng. Chem. Res. 2002, 41, 2379. [3] Van Speybroeck V. , Hemelsoet K. , Minner B. , Marin G. B. and Waroquier M. , Mol. Sim. 2007, accepted. [4] Hemelsoet K. , Moran D. , Van Speybroeck V. , Waroquier M. And Radom L., J. Phys. Chem. A 2006, 110, 8942. [5] Hemelsoet K. , Van Speybroeck V. , Moran D. Marin G. B. , Radom L. and Waroquier M. , J. Phys. Chem. A 2006, 110, 13624.

Web Page: molmod.ugent.be

632a Molecular-Based Modeling Of The Thermodynamic And Kinetic Costs Of Ion Desolvation (Regarding The Commercial Application Of Lithium-Ion Batteries)

Kevin L. Gering, Idaho National Laboratory, P. O. Box 1625, Idaho Falls, ID 83415-2209

Ion solvation figures centrally to properties of electrolytes tied to natural, industrial, commercial, and biological systems. The existence of ion solvation is a two-edged sword, enabling high salt dissociation, yet producing rate and energy barriers if ions are to be desolvated in the intended application. One such application is rechargeable lithium-ion (Li-ion) batteries, where lithium ions must generally become desolvated prior to intercalation into the electrode host materials. This cycle of ion solvation/desolvation occurs whenever such batteries are operated. This work will demonstrate that lithium desolvation is a process defined by successive removal of ligands from the ion solvation shell. There are time and energy requirements tied to this ligand-wise process, where the time and energy needed to remove successive solvent molecules from a solvated lithium ion become greater as we approach complete desolvation of the ion. It is believed that lithium ion desolvation contributes to poor performance of Li-ion batteries at low temperatures, since both the net time and energy required for desolvation increase at colder conditions. Thermodynamic and kinetic results for conventional battery electrolytes (such as ethylene carbonate + ethylmethyl carbonate + $LiPF_6$) will be shown, as simulated by a contemporary electrolytes code developed for use in the US DOE FreedomCAR program. The model views chemical physics per the Associative Mean Spherical Approximation, and receives other molecular-scale information from an "equation of state" that provides explicit treatment of ion solvation quantities. This modeling tool can be used to screen and optimize candidate electrolyte systems based on the requirements of minimizing solvent-ion interactions while maintaining favorable transport properties (electrical conductivity, viscosity, ionic diffusivity, etc.). Modeling results indicate that step-wise lithium desolvation is a strong function of lithium solvation number, and becomes more problematic at low temperatures and lower salt concentrations where lithium solvation numbers are higher. Lithium solvation numbers of two to five are common for electrolyte systems found in most Li-ion batteries. Lithium ion desolvation time and energy show similar trends over temperature and salt concentration, although they are obtained through independent approaches. While Li-ion batteries are used as a test case for this modeling capability, other applications abound in areas of industrial and biological systems.

Non-Equilibrium Molecular Dynamics Simulations of Yttria-Stabilized Zirconia Under an Alternating Electric Field

Kwong-Yu Chan and Qingyin Zhang

Department of Chemistry, The University of Hong Kong,

Pokfulam Road, Hong Kong.

Fax: 2857 1586; e-mail address: hrsccky@hku.hk

Abstract

Yttria-stabilized zirconia(YSZ) is a high-temperature electrolyte commonly used in solid oxide fuel cells. Electrochemical impedance spectroscopy (EIS) analyses have been routinely applied experimentally to probe the electrical conductivity of YSZ. In this presentation, an alternate current non-equilibrium molecular dynamics (AC-NEMD) simulation technique is used to study the oxygen anion conduction behavior of 8% mol yttria-stabilized zirconia(YSZ). With ohmic heat generated by the external field, a constant temperature iss effectively maintained by both the Anderson and Nose-Hoover thermostats. Both thermostats yield similar results. The simulation results indicate a resistance-capacitance (RC) circuit behavior at low frequency and resistance-inductance (RL) circuit behavior at high frequency. The transition between RC and RL behavior is connected with the time of oxygen ions transport between oxygen vacancies in YSZ.

Introduction

In recent years, YSZ has been intensively studied by molecular dynamics [1-14] simulations. Simulation research has focused on the self-diffusion coefficients (or conductivities) of oxygen ions, the activation energy of diffusion, and the structures of oxygen vacancies. Most simulations of solid oxide electrolytes reported were performed with the equilibrium molecular dynamics method. In EMD simulations, there is no net flux of oxygen ions, as opposed to the situation of a current flow in the operation of a solid-oxide electrolyte fuel cell. Only the self-diffusion of ions is calculated and ionic conductivity is evaluated by the Nernst-Einstein relationship between diffusion coefficient and conductivity. To simulate the situation of current flow, Tang et al. [15] first employed non-equilibrium molecular dynamics (NEDM) method with a direct current to study the conductivity of YSZ. In their results, the Arrhenius activation energy obtained by NEMD was in better agreement with experimental data compared to the results of EMD simulations. In their other simulation study of aqueous electrolyte confined in a nanopore, Tang et al. [16] observed a large discrepancy between EMD and NEMD results showing the Nernst-Einstein relationship to be not appropriate in a nano-confined system. The direct current non-equilibrium molecular dynamics (DC-NEMD) method is therefore more realistic than the EMD method in simulating current flow in a solid oxide electrolyte. In experiments, it is also very common to use of alternating current (AC) and impedance techniques to investigate electrode-electrolyte interfaces and electrolytes. Relaxation dynamics in multiple time scales can be probed by the application of AC and the analyses in frequency domain. The use of alternating current non-equilibrium molecular dynamics (AC-NEMD) method in simulating oxygen ion transport will be attractive and offers a new angle to probe the properties of the solid oxide electrolyte. The use of AC-NEMD has been applied to simulate an aqueous electrolyte confined in a nanopore [17,18]. The modes of ionic transport are markedly different in a solid-oxide electrolyte compared to aqueous electrolyte in which more than one ion species are mobile in the liquid phase. It will be useful to apply the AC-NEMD method to study the ionic transport behavior in YSZ.

Method

The 8 mol% YSZ has a cubic fluorite type structure with the cations forming a face-centered-cubic (FCC) arrangement. The simulation cell is composed of 3×3×3 FCC unit cells, containing 92 Zr^{4+}, 16 Y^{3+} and 208 O^{2-} ions. The lattice of zirconium (IV) oxide was first constructed and the positions of yttrium ions were generated by random replacements of zirconium ions. To keep the neutrality of simulation box, a corresponding number of oxygen ions were removed accordingly. A unit structure of 8 mol% YSZ is shown in Figure 1.

In the AC-NEND simulations, a sinusoidal electric field with a fixed frequency was imposed in the Z direction, the current response was monitored by tracing the velocity change of oxygen ions with time. The interaction between ions consists of short-range Born-Meyer-Buckingham (BMB) potential and the long-range Coulomb interaction potential. The BMB potential is expressed as

$$U(r_{ij}) = A_{ij} e^{-B_{ij} r_{ij}} - \frac{C_{ij}}{r_{ij}^6} \qquad (1)$$

where r_{ij} is the distance between the ions i and j; A_{ij}, B_{ij} and C_{ij} are the parameters of the interaction potential. The parameters used are identical to previously adopted values in literature [19, 20] and are given in Table 1. The Coulomb interaction has the form

$$U_c(r_{ij}) = \frac{q_i q_j}{4\pi\varepsilon_0 r_{ij}} \qquad (2)$$

where ε_0 is the permeability of free space, $q_i = z_i e$ and $q_j = z_i e$ are the charges of the ions. The charge valencies z_i, of zirconium, yttrium and oxide ions are +4, +3, and –2, respectively. Ewald summations [21] are used to calculate the long-range Coulomb interactions between ions in the simulation cell. Temperatures of the simulations here were at 900, 1273 and 1759K. The thermal expansion was accounted for by using the same number of ions but different sizes of the simulation cell. The length of the simulation cell, $a(T)$, increases with temperature and is given by the equation

$$a(T) = a(T = 300\text{K})[1 + \alpha(T - 300)] \qquad (3)$$

where $a(T = 300\text{K}) = 5.136$Å and $\alpha = 1\times10^{-5}$ K^{-1} is the thermal expansion coefficient for all the compositions of yttria-stabilized zirconium [3, 22].

The frequency-dependent alternating electric field $E(\omega)$ is applied along the Z direction of the simulation box . The equation of motion can be expressed as

$$\dot{\mathbf{q}}_i = \frac{\mathbf{p}_i}{m} \qquad (4a)$$

$$\dot{\mathbf{p}}_i = \mathbf{F}_i + q_i E_z(\omega)\hat{\mathbf{k}} \qquad (4b)$$

where the bold letters p_i and q_i are momentum and position vectors of ion i, m is its mass, F_i is the interaction force on the ion, q_i the charge of the ion, E_z is the strength of the applied field, ω is the frequency of external alternate current, and k is the unit vector in the same direction as the applied field. The current density of oxygen ions, $J_z(\omega)$, is calculated by

$$J_z(\omega) = \frac{1}{V}\sum_{i=1}^{N} q_i v_{i,z} \qquad (5)$$

where $v_{i,z}$ is the axial component of the velocity of ith oxygen ion, q_i is the valence charge of the ion and V is the volume of the simulation box. The relationship of Frequency dependent conductivity, $\sigma(\omega)$, current density, $J(\omega,)$ and electrical field, $E(\omega)$, can be expressed as

$$\boldsymbol{\sigma}(\omega) = \mathbf{J}(\omega)/\mathbf{E}(\omega) \qquad (6)$$

here, the bold face symbols denote complex numbers representing current density, electric field and conductivity. In simulations, the external AC electric field was applied in the Z direction.

The real functions E_z and J_z are the physical quantities implemented and measured in the NEMD simulations reported in this study.

In the linear response range, the complex conductivity can be obtained by equation (6) and resolved into real and imaginary parts, denoted by $\sigma'(\omega)$ and $\sigma''(\omega)$, respectively. The phase difference between external field and response current can be measured by fitting the current density with a sine function and comparing with the electric field function. Both Anderson [23] and Nose-Hoover [24, 25] thermostats were separately used. In the Anderson thermostat, the collision frequency ν was 0.00045 in reduced units similar to Tang et al [15]. In the Nose-Hoover thermostat, the parameter Q=0.5 in reduce units was used. The time step of simulation was set to 1 fs. Both thermostats controlled the desired temperature well.

Results

The NEMD simulations are performed at 900, 1273 and 1759K. The three different choices of amplitude for the external sinusoidal electric field are 4.47, 8.94, and 13.4×10^7 V m^{-1}. These correspond to reduced units of E* 1.0×10^{-6}, 2.0×10^{-6} and 3.0×10^{-6}, respectively. In experiments, a voltage of 1.0 can be applied over a thin film solid-electrolyte layer of 10^{-7}m so the electric field strengths simulated are not unrealistic values. The total run time of a simulation in each case is 1 ns, corresponding to 1 million time steps. The frequency of the sine electrical field ranges from 1.26×10^{13} Hz to 5.24×10^{13} Hz corresponding to reduced units of ω * = 0.01726 to 0.07193. These values are much large than reported experimental frequencies for YSZ impedance measurements.

At a given temperature, YSZ maintains a stable solid structure at all magnitudes and frequencies of electric fields applied, as shown in the pair distribution curves in Figure 2. The Zr-Zr and Zr-O pair distributions show pronounced isolated peaks, characteristic of a solid like structure. The oxygen oxygen pair has a lower first peak at about 0.26 nm and a broader and more connected second peak at about 0.37 nm. The oxygen distribution is more continuous, due to its larger number and the existence of vacancy sites. This continuous distribution also favors oxygen transport and almost all of the current is carried by the oxygen anions.

To check the size effect of simulation cell, additional simulations with a cell containing 4 ×4×4 FCC unit cells is employed for the case of 1759 K with $\omega^*=0.4316$. The comparison of current responses in simulations with a 4×4 ×4 cell and a 3×3×3 cell is shown in Figure 3. Both sets of simulations give very similar results. Comparing the applied potential waveform and the current response, phase shits can be calculated. At low frequency, the current response lags the electric field. At the intermediate frequency, the current response is in phase with the applied sinusoidal electric field, whereas at higher frequency, the current response leads the electric waveform. The phase shifts are shown graphically in Figure 3 for results over a range of temperature 900 to 1759 K.

Conductivity can be determined by equation (6) using the data of current responses at different field strengths. The conductivity determined assuming a linear response to a given field strength at a fixed frequency is shown in Figure 5. Fairly consistent values of conductivities are determined at different electric field strengths, indicating good linear response behavior and the results can be extrapolated to obtain the zero-field conductivity. The imaginary part of admittance is calculated in the same manner. The real part of admittance (conductivity) thus obtained is plotted in Figure 5 as a function of frequency. The electrical conductivity of YSZ has a maximum value at a certain frequency. This frequency corresponds to that of zero phase-shift in Figure 4. This maximum conductivity may be closely related to the optimum mode of oxygen transport between vacancy sites. In a given cycle of an applied waveform at a low frequency, an oxygen anion travels a large distance, beyond the next vacancy site. This can result in an ionic arrangement with increased polarity and undesirable configuration during field reversal. On the other hand, at a high frequency, an oxygen anion vibrates about its equilibrium position without being connected to the vacancy sites. Both of these situations do not carry much current. The maximum AC conductivity observed here is two orders of magnitude higher than the direct current conductivity observed experimentally and in previously DC-NEMD result [15]. The low frequency conductivity in Figure 6 does tend towards a very small number. It appears that the phenomenon of high frequency AC ionic conduction can be markedly different from the DC conduction. The effect of frequency on conductivity can also be interpreted by representation in the complex domain, indicating capacitance and inductance effects with negative and positive phase shifts, respectively. The complex representation is called admittance and Figure 6 plots the imaginary part versus the real part of admittance for the 8 mol% YSZ simulated at three different temperatures. This σ'-σ'' plot or Cole-Cole plot is analogous to the Z′-Z" plot of impedance in the complex domain, when impedance is the reciprocal of admittance. The plot of complex impedance is a popular experimental technique called impedance spectroscopy (IS) [26]. In Figure 6, admittance starts at low frequency in the fourth quadrant with negative complex component indicating the presence of a capacitance. At an intermediate frequency, the maximum conductivity (real part) is reached with a zero phase shift when the curve crosses the x-axis. At the highest frequencies, the admittance curve resides in the first quadrant and an inductance component is indicated. The admittance approaches the origin at the infinite frequency limit. Most experimental impedance spectroscopy data show only a semi-circle in the first quadrant, which is equivalent to a semi-circle of admittance in the fourth quadrant of the Cole-Cole plot. The equivalent circuit of YSZ and other solid-state electrolytes interpreted from experimental IS are dominated by resistance-capacitance (RC) components without any inductance-resistance

(LR) components. A number of factors can contribute this difference between the AC-NEMD results here and experimental results. There is a big gap between the lowest AC frequency that can be simulated and the highest frequency applied in experiments. Recent experiments also focused more on the performance of anode or cathode and the electrode-electrolyte interface whereas our present NEMD study is a uniform bulk YSZ material without electrode-electrolyte interfaces. Another important difference is the presence of grain boundaries in YSZ bulk material studied experimentally. The relaxation frequency of bulk lattice is one or two orders of magnitude higher that in between grain boundaries. These differences make it difficult to have a direct comparison of the present NEMD results and experimental data. Nevertheless, the AC-NEMD simulation presented here is a useful step towards the comprehensive modeling of solid oxide electrolyte at the atomic scale.

Acknowledgement

This research has been supported by a Hong Kong Research Grants Council grant (HKU 7006/05P).

Table 1 Inter atomic potential parameters.

$i-j$	A_{ij} (kJ mol^{-1})	B_{ij} (m^{-1})	C_{ij} (kJ nm^6 mol^{-1})
Zr – Zr	0	1	0
Zr –Y	0	1	0
Y – Y	0	1	0
Zr – O	9.512×10^4	2.660×10^{10}	0
Y – O	12.978×10^4	2.865×10^{10}	0
O – O	2.196×10^6	6.711×10^{10}	2.691×10^{-3}

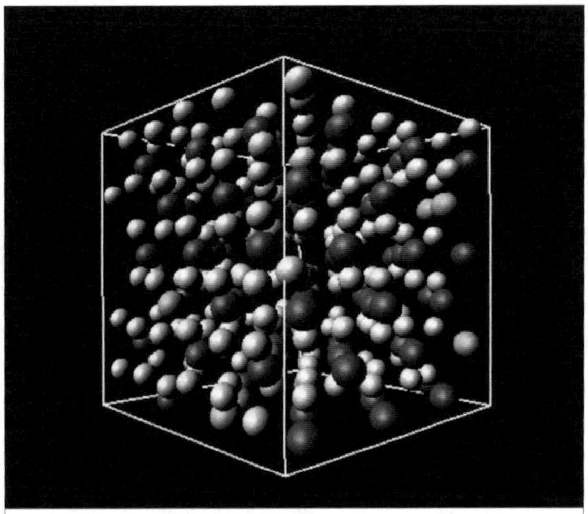

Fig. 1. Setup of NEMD: cubic lattice of ZrO_2 with some Zr^{4+} replaced by Y^{3+}.

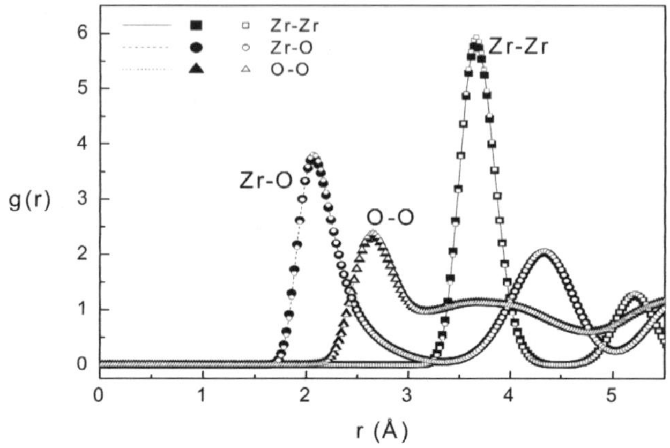

Figure 2. Radial distribution functions of three different atom pairs in 8% YSZ at 1759 K in an AC electric field of strength $E^* = 3 \times 10^{-6}$. (lines: $\omega^* = 0.07193$, solid symbols: $\omega^* = 0.04316$, open symbols: $\omega^* = 0.01726$)

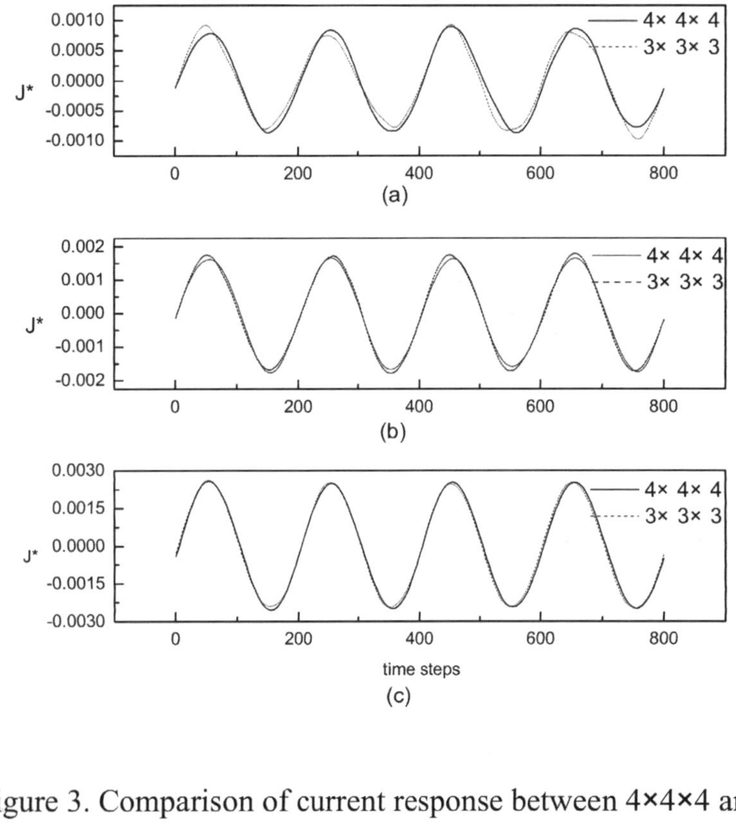

Figure 3. Comparison of current response between 4×4×4 and 3×3×3 cell under an Anderson thermostat. (Increasing electric field strengths from top to bottom: $E^* = 1\times10^{-6}$, 2×10^{-6}, 3×10^{-6})

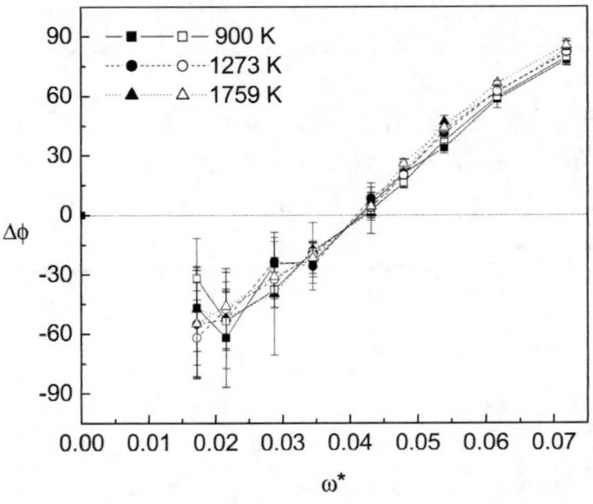

Fig. 4. Phase shift of currents in 8% YSZ versus frequency of electric field at different temperatures (solid symbols: $E^*=1\times10^6$; open symbols $E^*=3\times10^6$; error bars are statistical deviations in a simulation run).

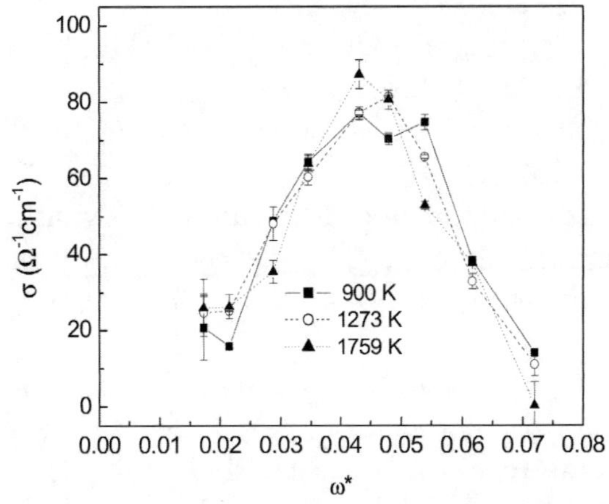

Figure 5. Frequency dependence conductivity of 8% YSZ simulated with an Anderson thermostat.

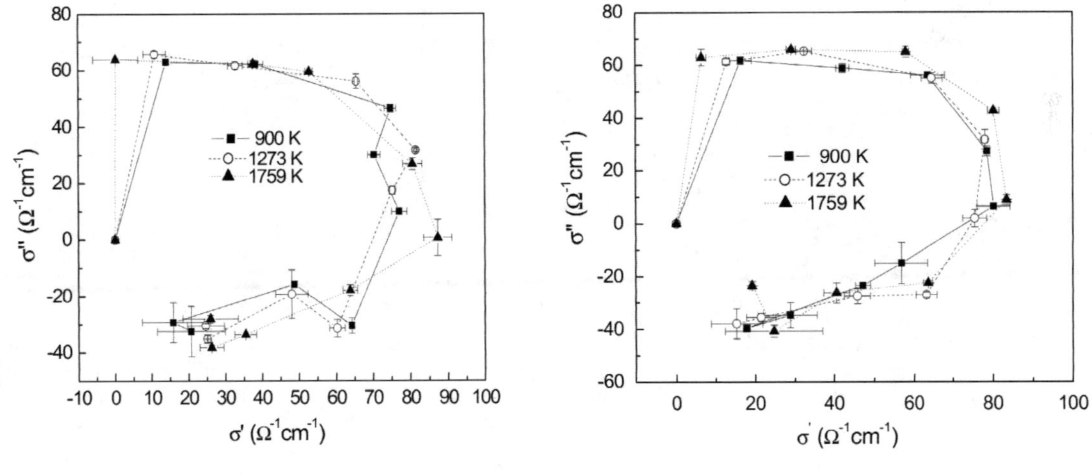

Figure 6. Cole-Cole plot of frequency dependent admittance in 8% YSZ simulated with an Anderson thermostat (left panel) and Nose-Noover (right panel) thermostats.

References:

[1] Shimojo, F.; Okabe, T.; Tachibana, F.; Kobayashi, M.; Okazaki, H. *J. Phys. Soc. Jap.* **1992**, 61, 2848

[2] Brinkman, H. W.; Briels, W. J.; Verweij, H. *Chem. Phys. Lett.* **1995**, 247, 386

[3] Li, X. Y.; Hafskjold, B. *J. Phys.Cond. Matt.*

1995, 7,1255
[4] Khan, M. S.; Islam, M. S.; Bates, D. R. *J. Mater. Chem.* **1998**, 8, 2299
[5] Yamamura, Y.; Kawasaki, S.; Sakai, H. *Solid State Ion.* **1999**, 126, 181
[6] Fisher, C. A. J.; Matsubara, H. *Solid State Ion* **1998**, 115, 311
[7] Suzuki, K.; Kubo, M.; Oumi, Y.; Miura, R.; Takaba, H.; Fahmi, A.; Chatterjee, A.; Teraishi, K.; Miyamoto, A. *Appl. Phys. Lett.* **1998**, 73, 1502.
[8] Sawaguchi, N.; Ogawa, H., Simulated diffusion of oxide ions in YO1.5-ZrO2 at high temperature. *Solid State Ion* **2000**, 128, 183
[9] Perumal, T. P.; Sridhar, V.; Murthy, K. P. N.; Easwarakumar, K. S.; Ramasamy, S. *Physica. A* **2002**, 309, 35
[10] Kilo, M.; Argirusis, C.; Borchardt, G.; Jackson, R. A. *Phys. Chem. Chem. Phys.* 2003, 5, 2219
[11] Arima, T.; Fukuyo, K.; Idemitsu, K.; Inagaki, Y. J. *Mol. Liq.* **2004**, 113, 67
[12] Kilo, M.; Taylor, M. A.; Argirusis, C.; Borchardt, G.; Jackson, R. A.; Schulz, O.; Martin, M.; Weller, M. *Solid State Ion.* **2004**, 175, 823
[13] Hayashi, H.; Saitou, T.; Maruyama, N.; Inaba, H.; Kawamura, K.; Mori, M. *Solid State Ion.* **2005**, 176, 613
[14] Devanathan, R.; Weber, W. J.; Singhal, S. C.; Gale, J. D. *Solid State Ion.* **2006**, 177,1251
[15] Tang, Y W.; Zhang, Q. Y; Chan K.Y. *Chem. Phys. Lett.* **2004**, 385, 202.
[16] Tang, Y W.; Szalai, I.; Chan K.Y. *J. Phys. Chem. A.* **2001**, 105, 9616
[17] Tang, Y. W.; Chan, K. Y.; Szalai, I. *Nano. Lett.* **2003**, *3, 217*
[18] Tang, Y. W.; Szalai, I.; Chan, K. Y. *Mole. Phys* **2002**, 100, 1497
[19] Lewis, G. V.; Catlow, C. R. A. *J. Phys. C* **1985**, 18, 1149
[20] Dwivedi, A.; Cormack, A. N. *Phil. Mag. A,* **1990**, 61, 1
[21] Ewald, P P. *Ann. Phys.* **1921**, 64, 253
[22] Ingel, R. P.; Lewis, D. *J.Am.Ceram.Soc.* **1986,** 69, 325
[23] Andersen, H. C. *J. Chem. Phys.* **1980**, 72, 2384
[24] Nose, S. *J. Chem. Phys.* **1984**, 81, 511
[25] Hoover, W. G. *Phys. Rev. A* **1985**, 31, 1695
[26] Barsoukov, E.; Macdonald, J. R. *Impedance Spectroscopy, Theory, Experiment, and Applications.* **2005**, John Wiley & Sons, Inc., Hoboken, New Jersey.

632c Stability And Sorption Behavior Of Thiamine Hydrochloride Hydrates

Neeraj Rai[1], Paroma Chakravarty[2], Raj Suryanarayanan[2], and J. I. Siepmann[1]. (1) University of Minnesota, Department of Chemistry, Minneapolis, MN 55455-0431, (2) Department of Pharmaceutics, University of Minnesota, Minneapolis, MN 55455

Monte Carlo simulations in isotension-isothermal and Gibbs ensemble are carried out to investigate the stability and sorption behavior of the nonstoichiometric hydrate (NSH), isomorphic desolvate (ID), and hemihydrate (HH) of thiamine hydrochloride at different temperatures and relative humidities. The TraPPE and TIP4P force fields are used to model thiamine hydrochloride and water, respectively. The simulations provide molecular-level understanding on the different sorption behavior of these pharmaceutical polymorphs.

632d Understanding Molecular Interactions In Organic-Sulfur Dioxide And Organic-Mercury Systems

Abhijit V. Shevade[1], Amy Ryan[1], Margie Homer[1], Mario Blanco[2], and William A. Goddard III[2]. (1) Jet Propulsion Laboratory, 4800 Oak Grove Drive, M/S 184-105, Pasadena, CA 91109, (2) Materials and Process Simulation Center (MSC), California Institute of Technology, Beckman Institute (139-74), Pasadena, CA 91125

The objective of this investigation is to use first principles calculations to understand the gas phase interactions of organic molecules with sulfur dioxide (SO_2) and elemental mercury (Hg^o). Developing an understanding of these systems is of great interest for environmental monitoring and remediation applications, as several chemical sensing systems use organic materials as sensing elements.

Common classes of organic structures, including alkanes, alkenes, aromatics, primary, secondary and tertiary amines, amides, aldehydes, and carboxylic acids, were studied. Binding energies between SO_2, Hg^o and each organic molecule were calculated quantum mechanically (QM), using B3LYP and X3LYP flavors of Density Functional Theory (DFT). In addition, scans of the potential energy surfaces between these various organic molecules and SO_2 and Hg^o were performed, and the van der Waals interactions were fit to develop a first principles force field (Morse potential), for use in molecular dynamics (MD) simulations. The forcefield parameters were calculated for various atom types in the organic molecules, such as carbon (sp^3, sp^2, sp hybridized), nitrogen (sp^3, sp), oxygen (sp^2), for example; for their interactions with the sulfur, oxygen and mercury atoms in SO_2, Hg^o.

The QM binding energy results for the organic-SO_2, Hg^o systems were analyzed for their van der Waals and electrostatic energy contributions. The calculated QM binding energies for organic-SO_2 systems indicate that the organics that show strong binding with SO_2 are ones containing amine, amide and acid groups. Weak or no binding of SO_2 was observed with alkane, alkene, and aromatics. The binding energies for organic-Hg^o systems showed that the only organics that had any binding are primary and secondary amines, which showed only weak binding. All other organic molecules showed repulsive interactions with Hg^o.

The organic molecules considered in this study are considered to be a part of a polymer chain, either in the backbone or as side groups. The methodology has been successfully used for rapid quantum screening of polymer based sensing materials in the JPL Electronic Nose (ENose).

Keywords: Quantum mechanics, Density Functional Theory, Binding Energy, Organics, Mercury, Sulfur dioxide, Computational material screening, Environmental monitoring

Density Functional Theory Characterization of Zinc Orthotitanate Sorbents for Hydrogen Sulfide Dissociation

Rees B. Rankin[1], Karl Johnson[2], David S. Sholl[3] and Shiqiang Hao[3], (1)Department of Chemical and Petroleum Engineering, University of Pittsburgh, Benedum Hall #1123, University of Pittsburgh, Pittsburgh, PA 15261, (2)University of Pittsburgh, 1249 Benedum Hall, Department of Chemical Engineering, Pittsburgh, PA 15261, (3)Carnegie Mellon University, Department of Chemical Engineering, Carnegie Mellon University, Pittsburgh, PA 15213

Integrated Gas Combined Cycle (IGCC) technology provides a method to produce electricity from coal in a clean, efficient way. One of the first steps in the IGCC process is the production of syngas from coal, which must be cleaned before being used in a gas turbine. A low cost high uptake sorbent is needed to capture contaminant species prior to the turbine entrance. Zn_2TiO_4 ('ZTO') has been proposed as this sorbent. ZTO has been identified, in part for its high performance in removal of gas phase H_2S. H_2S is one of the main contaminants in this syngas stream. To help quantify the performance of ZTO as a sorbent for H_2S at the atomic scale, we have used first principles density functional theory (DFT) calculations to examine the dissociative adsorption of H_2S on ZTO.

We first calculated the low energy bulk structure(s) of the inverse spinel ZTO with DFT. DFT calculations were used to identify the ground state structure of bulk ZTO because it was not conclusively identified via previous experimental analysis.[1] In these prior experiments the configurational occupation of the sublattice octahedral sites with Zn and Ti cations could not be determined. The energetic distribution associated with changes in the octahedral site configurations was examined with 20 distinct DFT calculations. The results of these DFT calculations were used to build a predictive Cluster Expansion model for the energy of any arbitrary ZTO bulk structure. The Cluster Expansion model was also used in conjunction with MC simulations to obtain a profile of temperature dependence on the material bulk phase distribution. The MC simulations revealed that the ground state bulk structure of ZTO should be the significant dominant phase in real ZTO materials up to temperatures of approximately 900K. The ground state structure of ZTO identified via DFT calculations is shown in Figure 1.

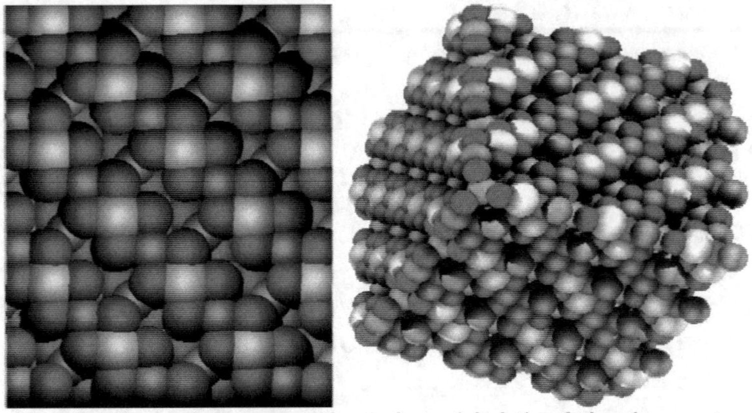

Figure 1: A face on view (left) and perspective view (right) of the lowest energy bulk structure of ZTO identified from ZTO calculations are shown.

Using the identified low energy structure of bulk ZTO, we constructed numerous representative low Miller index surfaces of ZTO. Because bulk ZTO has anisotropic features, each Miller index surface ZTO can possess multiple distinct terminations. We calculated the surface energy of 40 surface terminations in the low Miller index families of ZTO surfaces using DFT. Our DFT calculations revealed that a termination of the ZTO(010) surface is the lowest energy surface that ZTO likely exposes. A slightly higher energy ZTO(110) surface was also identified. Figure 2 shows a schematic top down view of these two low energy ZTO surface terminations.

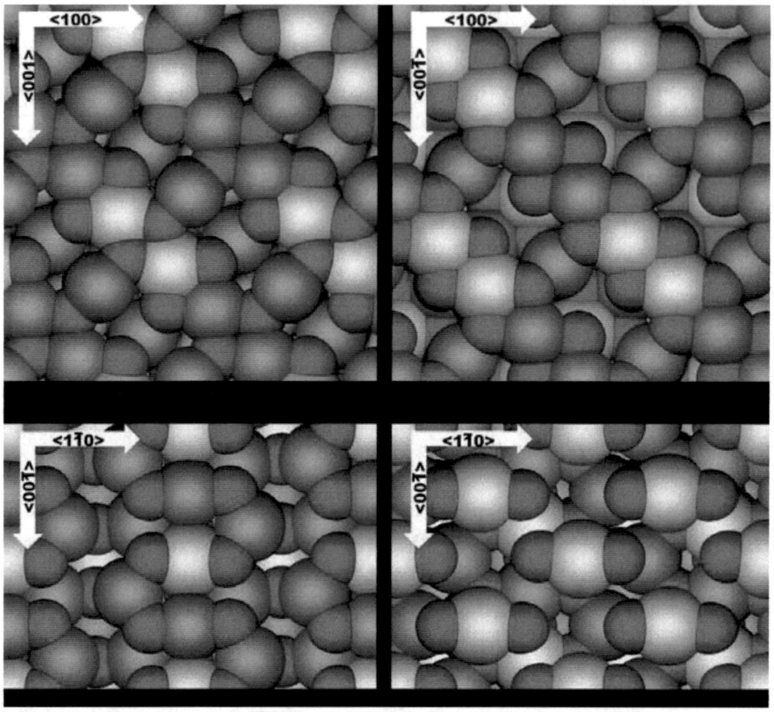

Figure 2: The ZTO(010) (top) and ZTO(110) (bottom) surface terminations with the lowest energy are shown. For each surface, the top and bottom sides of the calculation slab are shown.

With DFT calculations, we then examined H, S, and H_2S adsorption at low coverage on the low energy ZTO(010) surface. Atomic S was calculated to possess the lowest energy when bound to Zn-Zn bridge sites on the ZTO(010) surface. Atomic H was calculated to possess the lowest energy when bound to O-O bridge site in the ZTO(010) surface octahedral trench. The O-H bond length was found to be 1.0Å, supported by a secondary H-bonding chain to the opposing O atom in the O-O bridge site. A Zn-Zn bridge site was also determined to be the lowest energy binding site for H_2S. We then used these results in conjunction with the nudged elastic band method implemented in DFT calculations to probe the decomposition pathway of H_2S to 2H+S on the ZTO(010) surface. An initial reaction barrier of approximately 0.05 eV was calculated for the 1st dehydrogenation step, and a barrier of approximately 0.30 eV was calculated for the 2nd dehydrogenation step.

References:

[1] Bartram S F and Slepetys R A 1961 *J. Am. Ceram. Soc.* **44** 49

ADSORPTION STATES OF AMPHIPATIC SOLUTES AT THE SURFACE OF NATURALLY HYDROPHOBIC MINERALS: A MOLECULAR DYNAMICS SIMULATION STUDY

Hao Du and J. D. Miller*

Department of Metallurgical Engineering, 135 S. 1460 E., 412 William C. Browning Building. University of Utah, Salt Lake City, Utah 84112, USA

*Corresponding author E-mail: Jan.Miller@utah.edu

Keywords: Molecular dynamics simulation, talc, graphite, sulfur, DTAB, dextrin, adsorption state, hydrophobic interaction.

Naturally hydrophobic minerals such as graphite and talc are common gangue minerals found in sulfide ores and are difficult to separate due to their tendency to float together with valuable sulfide minerals. Despite the relatively successful use of the polysaccharide group of chemicals (dextrin, guar gum, carboxymethyl celluloes (CMC), etc,) as flotation depressants for these naturally hydrophobic minerals, the nature of the adsorption processes remains in debate. Consequently, the adsorption of amphipatic solutes at naturally hydrophobic minerals such as coal/graphite, talc, and sulfur is of interest to many researchers and substantial research has been reported[1-16].

In this study, the adsorption states of selected amphipatic solutes at selected naturally hydrophobic mineral surfaces were examined using MDS in order to provide further information regarding adsorption processes, especially with respect to the significance of hydrophobic interactions. Initially, the adsorption of a cationic surfactant dodecyltrimethylammonium bromide (DTAB) at a graphite surface was simulated since this adsorption process has been well recognized to involve hydrophobic interactions. In this way the effectiveness of MDS methodology for the study of surfactant adsorption can be assessed. Then, the behavior of DTAB at a hydrophobic talc basal plane surface was studied, and the significance of hydrophobic examined for this case.

Finally, for the adsorption of polysaccharide polymers, the behavior of a model dextrin molecule at graphite, talc, and sulfur surfaces was investigated. The different adsorption states of dextrin at these surfaces are discussed in terms of the molecular structure of dextrin, mineral surface hydrophobicity, surface atom polarity, and surface nanoroughness. Finally, some influential factors affecting the adsorption of dextrin at naturally hydrophobic mineral surfaces are summarized and discussed.

DTAB at the Graphite Surface

The DTAB adsorption process at a hydrophobic graphite surface is described in Figure 1. Initially, 24 DTAB ions are sandwiched between graphite surfaces with their polar head groups facing the substrates (Figure 1A). Due to the strong hydrophobic interaction between the substrates and the DTAB tails, spontaneous reorientation of surfactant molecules is observed as seen in Figure 1B and Figure 1C. The alkyl chain of the DTAB ion spreads at the graphite surface and the polar amine head groups stretch out into the aqueous phase, leading to a hydrophilic state at the graphite surfaces. Also, a spherical micelle structure appears to be formed in the bulk aqueous phase as observed early in the dynamics of the adsorption process (Figure 1B). Eventually, complete adsorption of the DTAB surfactant at the graphite surfaces is achieved (Figure 1D), suggesting a strong affinity of the DTAB molecules for the graphite substrate.

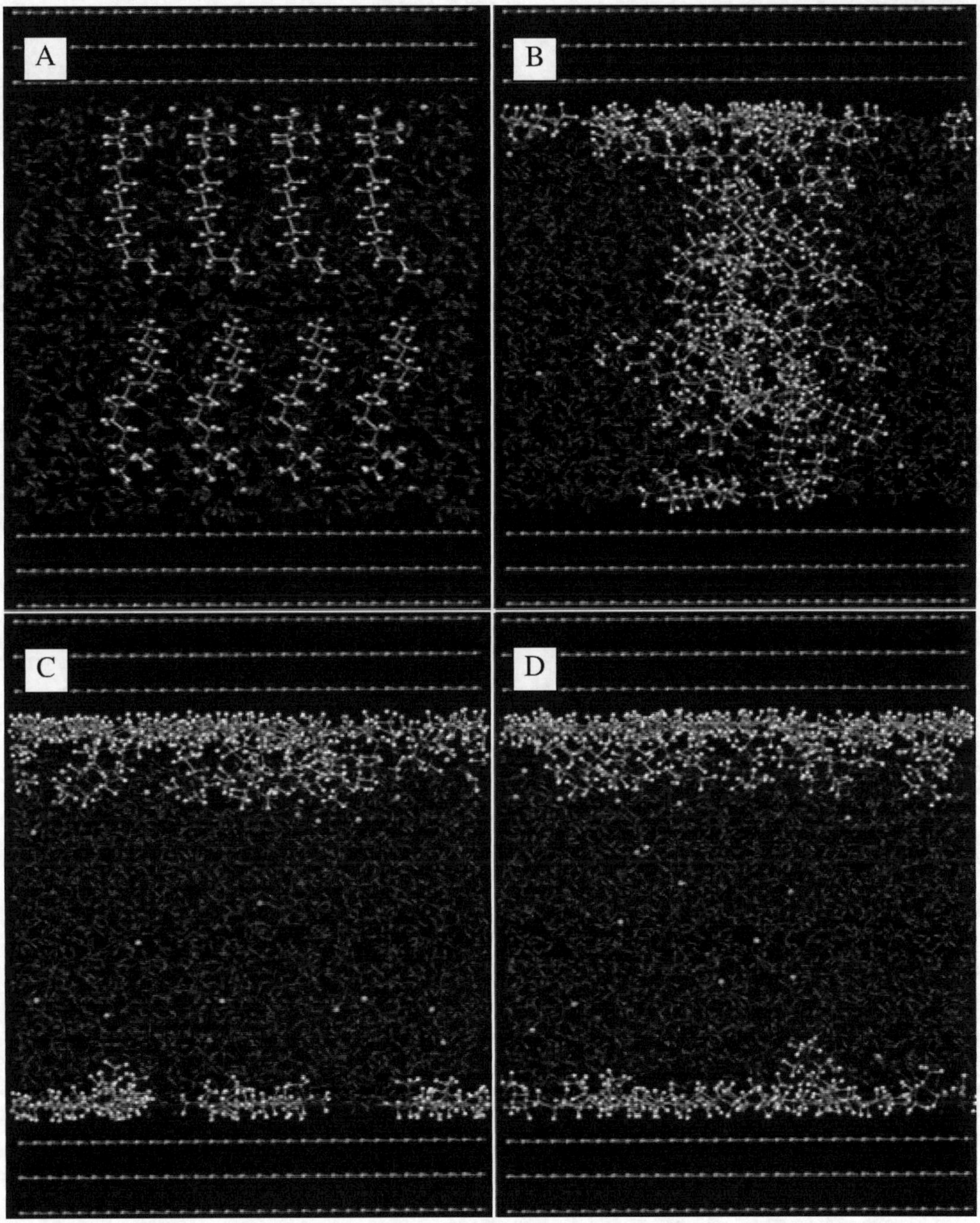

Figure 1. MDS snapshots of 24 DTAB molecules adsorption states at graphite surface for different simulation times. The snapshots are taken at 0 ps, 200ps, 500ps and 1ns for A, B, C, and D respectively. The color representations are as follows: Red-oxygen atoms, white-hydrogen atoms, blue-nitrogen atoms, green-bromide atoms, and light blue-carbon atoms.

DTAB at the Talc Basal Plane Surface

Snapshots of DTAB surfactant ions at the talc basal plane surface (001) are presented in Figure 2 for different simulation times. The initial configuration of DTABs was in a way that the polar head groups were facing the talc surface. Similar to DTAB/graphite system, after a short period of simulation, the surfactant molecules reorient themselves so that their hydrophobic tails move towards the surface, and displace water molecules as clearly represented in Figure 2B. Our previous study regarding water structure at talc basal plane surfaces has suggested that due to the absence of specific hydrogen bonding donor and/or acceptor sites on the basal plane of talc; the interaction between water molecules and the basal plane is weak[17,18]. The relatively weak water/talc interaction on the molecular scale is the origin of the macroscopic hydrophobic character of the basal plane surface. The observed spontaneous reorientation of DTAB ions is due to the hydrophobic nature of the talc basal plane, which favors the adsorption of the DTAB hydrophobic tails and the displacement of interfacial water molecules[17].

As time evolves, the hydrophobic interaction between the substrate and the tail of the surfactant is so dominating that almost all water molecules are excluded between them as shown in Figures 2C and 2D. Further study of the surfactants in Figures 2C and 2D reveals that at equilibrium, the DTABs form a hemi-spherical like micelle structure with the hydrophobic tails sticking to the substrate and the polar heads in contact with surrounding water molecules. Due to the orientation of adsorbed DTAB ions, a hydrophilic surface is produced.

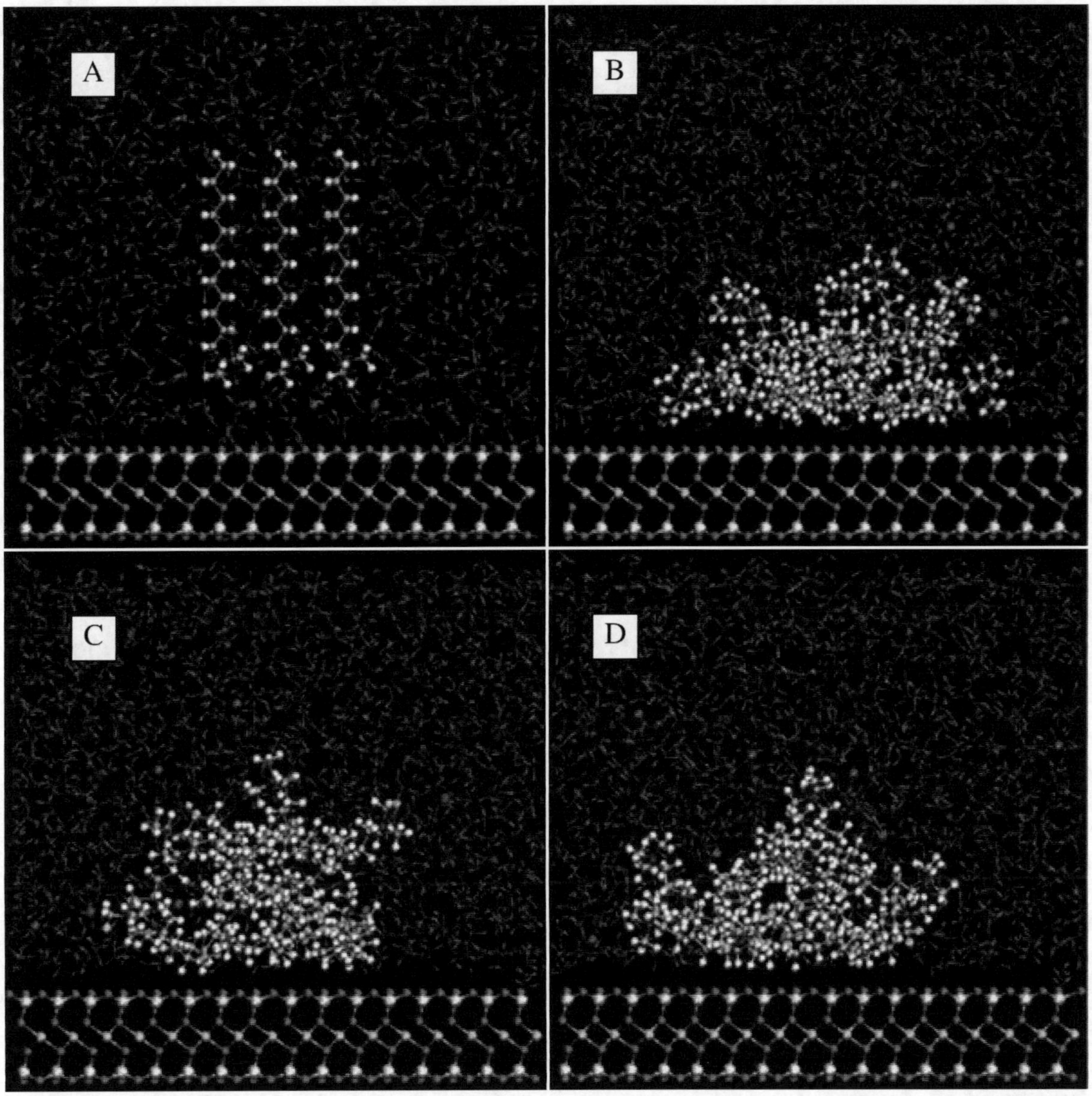

Figure 2. MDS snapshots of 9 DTAB molecules adsorption states at talc basal plane surface for different simulation times. The snapshots are taken at 0 ps, 200ps, 500ps and 1ns for A, B, C, and D respectively. The color representations are as follows: Red-oxygen atoms, white-hydrogen atoms, blue-nitrogen atoms, purple-bromide atoms, yellow-silicon atoms, green-magnesium atoms, and light blue-carbon atoms.

Dextrin at the Graphite Surface

The behavior of the dextrin molecule at the graphite surface is presented in Figure 3. It is noticed that, firstly, the "dextrin" molecule remains attached to the graphite surface at one end initially (From Figure 3A to Figure 3B), indicating its strong affinity to the graphite surface. Secondly, the "dextrin" molecule reaches full contact with the graphite surface as time evolves (Figures 3B, 3C, and 3D); suggesting that a complete spreading or accommodation of the "dextrin" molecule at the graphite surface is an energetically stable state.

In order to understand the "dextrin" organization at the graphite surface, the properties of the graphite surface as well as the structure of the dextrin molecule have to be considered. As mentioned previously, the graphite, composed of inactive neutral carbon atoms, is naturally hydrophobic and is not able to hydrogen bond with water molecules. Consequently, water molecules are excluded from the graphite surface. The dextrin molecule on the other hand, shows a dual characteristic in that the presence of hydroxyl groups provides this molecule with plenty of hydrogen bonding sites, and enables the dextrin molecules to hydrogen bond with surrounding water molecules. At the same time the low polarity of the methylene groups as well as the C-H groups in the "dextrin" molecule show a hydrophobic character, and these groups drive the molecule to escape from the water phase. When interacting with the hydrophobic graphite surface, the hydrophobic moieties of the "dextrin" molecule preferentially stay in contact with the graphite surface as suggested by Figure 3, which when compared to hydroxyl groups, the methylene and the C-H groups in the "dextrin" molecule are significantly closer to the graphite surface. The hydrophilic hydroxyl groups favor hydrogen bonding with water molecules. As a compromise, the "dextrin" molecule reorients to expose as many hydrophobic moieties as possible to the hydrophobic graphite surface while the majority of the dextrin hydroxyl groups hydrogen bond with water molecules. In conclusion, part of the "dextrin" molecule adsorbs at the graphite surface due to hydrophobic attraction, while the remainder of the molecule remains hydrated by the water phase through hydroxyl-water hydrogen bonding as observed in Figure 3. The ability of the "dextrin" molecule to be adsorbed at a hydrophobic surface through hydrophobic interactions and at the same time to form a hydrogen bonding network with water molecules in the aqueous phase enables it to behave like a bridge connecting the hydrophobic graphite surface with surrounding water molecules. In this way the naturally hydrophobic graphite surface is made hydrophilic.

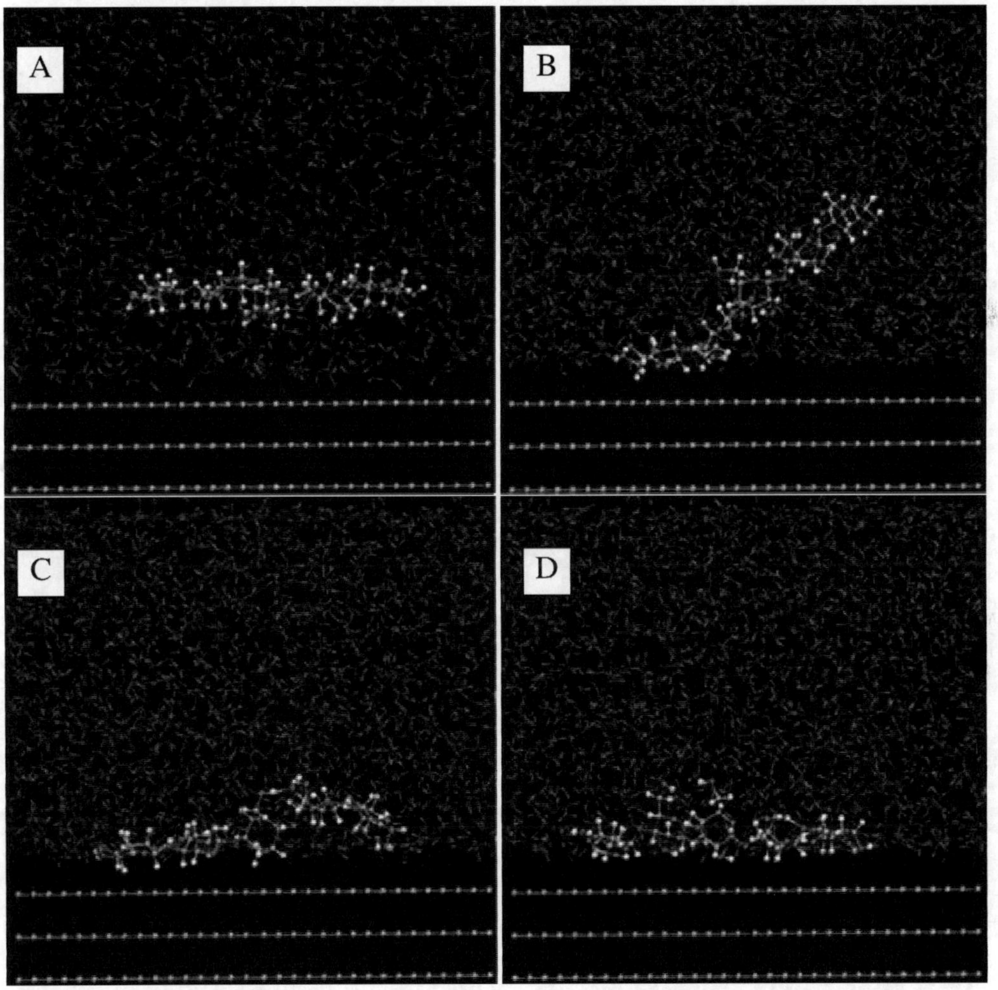

Figure 3. MD simulation snapshots of the "dextrin" molecule (5 dextrose monomers) adsorption states at graphite surface for different simulation times. The snapshots are taken at 0 ps, 200ps, 500ps and 1ns for A, B, C, and D respectively. The color representations are as follows: Red-oxygen atoms, white-hydrogen atoms, and light blue-carbon atoms.

Dextrin at the Talc Basal Plane Surface

The simulation of a "dextrin" molecule at the talc basal plane surface is presented in Figure 4. It is noticed that the adsorption state is very similar to that observed at a graphite surface. Initially, the "dextrin" molecules are positioned in the middle of the solution (Figure 4A). As the simulation time evolves, the "dextrin" organization migrates towards the talc basal plane surface (Figure 4A and Figure 4B), and remains in full contact with the talc surface once it reaches such a position (Figure 4C, and Figure 4D), indicating its strong affinity for the basal plane. As previously discussed, because of the absence of polarity, the talc basal plane surface is hydrophobic, thus hydrophobic interaction is proposed to be one of the major reasons for the "dextrin" adsorption[1,5,10]. On the other hand, despite the fact that the talc basal plane surface oxygen atoms are not ideal hydrogen bonding sites, because of the negative charge these oxygen atoms carry, they can weakly hydrogen bond with water molecules. Thus, compared to the carbon atoms at the graphite surface, these tetrahedral oxygen atoms at the talc surface are able to accommodate hydroxyl groups in the "dextrin" molecule to some extent. Consequently, both the hydrophobic moieties and the hydroxyl groups are stabilized at the talc basal plane surface, as suggested by Figure 4C and Figure 4D which show that both C-H/methylene and hydroxyl groups are very close to the substrate. In summary, hydrophobic interactions as well as hydrogen bonding between the talc basal plane surface oxygen atoms and hydrogen atoms in the hydroxyl groups contribute to "dextrin" adsorption at the talc basal plane surface.

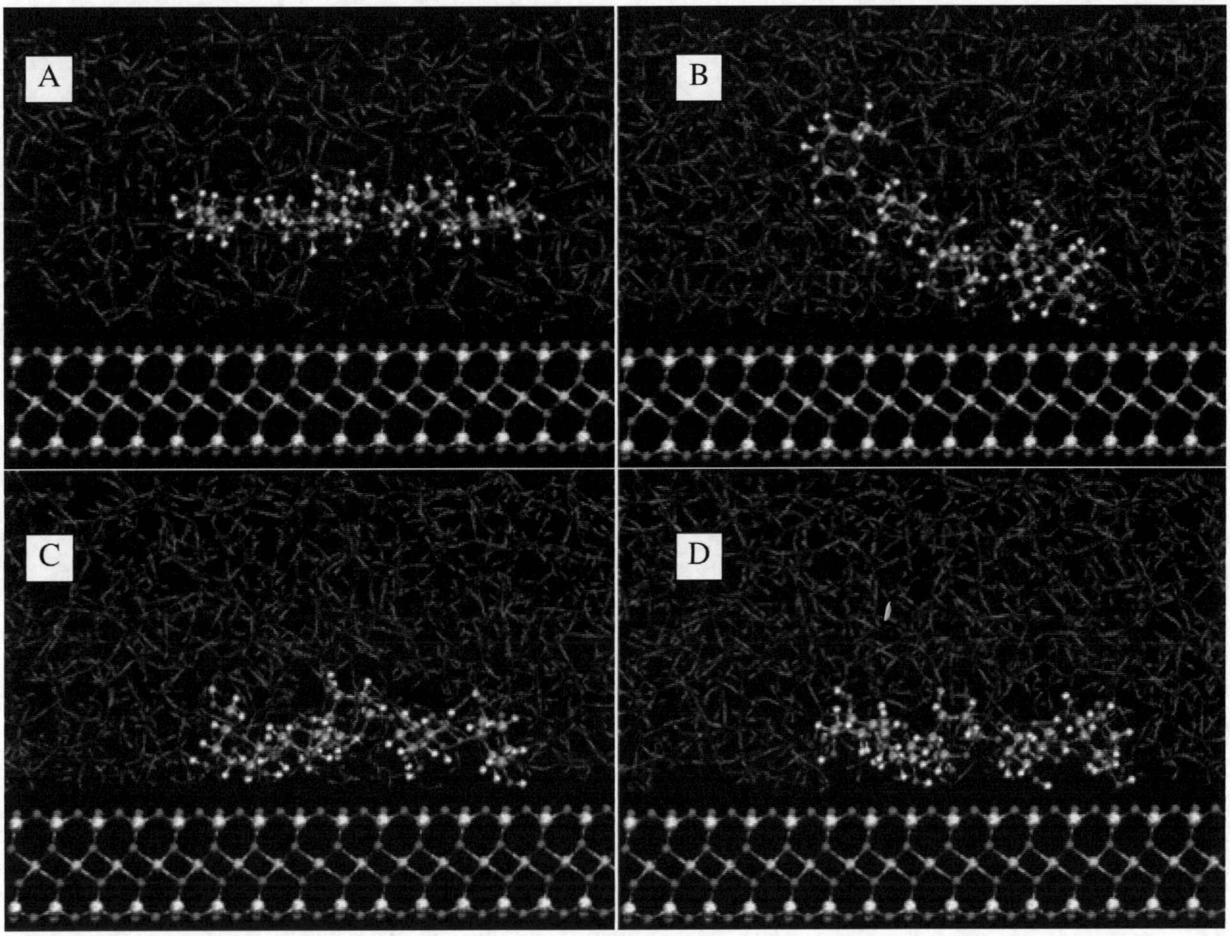

Figure 4. MD simulation snapshots of the "dextrin" molecule (5 dextrose monomers) adsorption states at talc basal plane surface for different simulation times. The snap shots are taken at 0 ps, 200ps, 500ps and 1ns for A, B, C, and D respectively. The color representations are as follows: Red-oxygen atoms, white-hydrogen atoms, yellow-silicon atoms, green-magnesium atoms, and light blue-carbon atoms.

Dextrin at Sulfur Surface

The adsorption of dextrin molecules by sulfur has been reported to be similar to that observed for talc and coal[16]. Hydrophobic interaction has been proposed to be the major adsorption mechanism[16]. In this study, a rhombic sulfur mineral has been cut alone the (111) plane to keep the integrity of each S_8 molecule. Our simulation results as illustrated in Figure 5 suggest that the "dextrin" molecule behaves very different from what it does at the graphite and talc basal plane surfaces. The adsorption of the "dextrin" molecule at the sulfur surface is achieved by the attachment of a portion of the molecule, and a complete adsorption has not been observed during the entire simulation time (Figure 5B, 5C, and 5D).

Due to the similarity of the sulfur and graphite substrate, which are composed of nonpolar atoms, the significantly weakened adsorption state is understood by considering the structural difference between the sulfur and graphite surfaces. It is known that rhombic sulfur is built up from cyclooctasulfur (S_8) molecules. Unlike graphite, which can be cleaved perfectly along its major cleavage plane (0001), the cleavage of a rhombic sulfur crystal is not able to produce an atomically flat surface due to its unique structure. A close examination of the sulfur surface (111) used in the simulation shows that the surface sulfur atoms are not positioned in the same plane. As a consequence atomic notches occur between each row of S_8 molecules as illustrated in Figure 5. This atomic nanoroughness provides structural barriers to the adsorption of "dextrin" hydrophobic moieties at the surface while simultaneously remaining in an energetically stable molecular configuration. On the other hand, small water molecules are capable of dynamically occupying these surface voids, thus provides additional energy barriers for the "dextrin" adsorption.

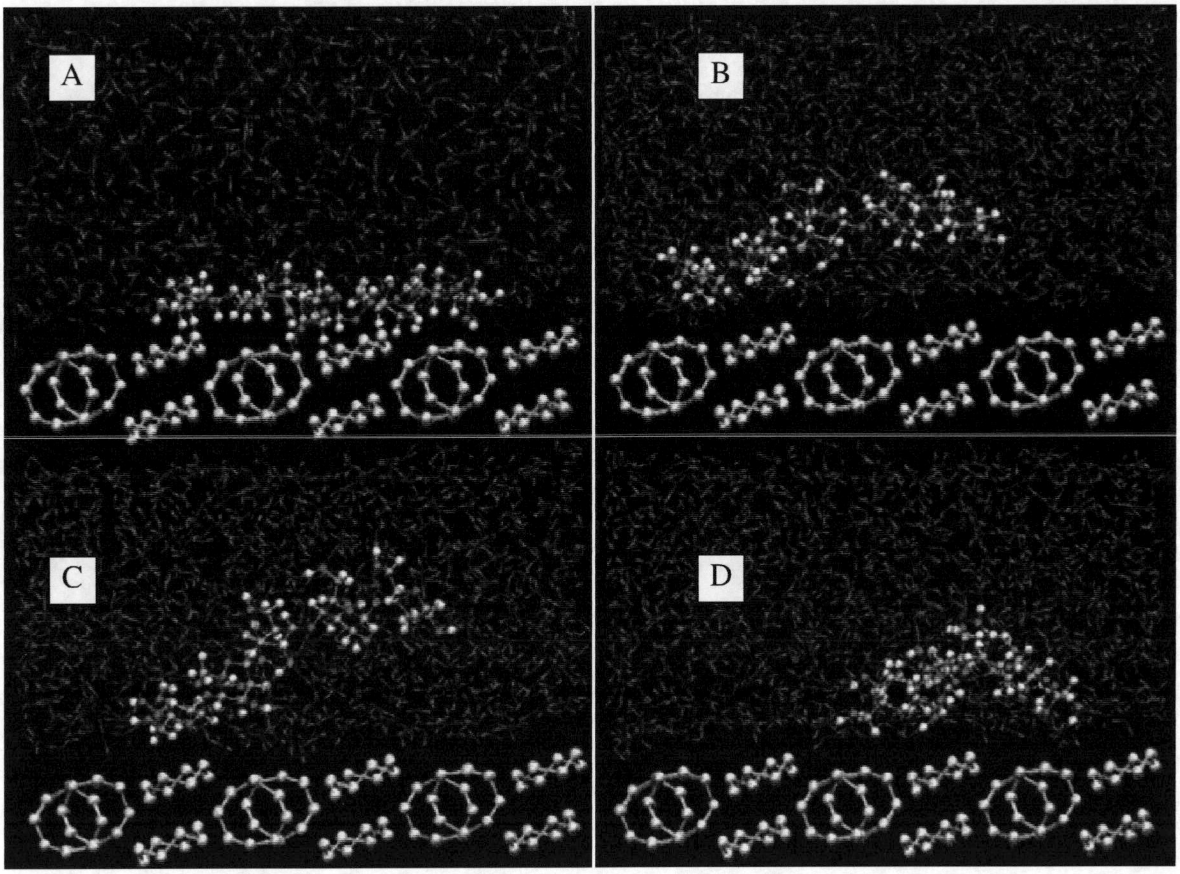

Figure 5. MD simulation snapshots of the "dextrin" molecule (5 dextrose monomers) adsorption states at sulfur (111) surface for different simulation times. The snap shots are taken at 0 ps, 200ps, 500ps and 1ns for A, B, C, and D respectively. The color representations are as follows: Red-oxygen atoms, white-hydrogen atoms, yellow-sulfur atoms, and light blue-carbon atoms.

Further Discussion

The MD simulation results regarding adsorption states of dextrin at graphite, talc, and sulfur surfaces have suggested that there are several important factors which influence the adsorption processes.

a) Naturally hydrophobic surfaces. The hydrophobic interaction between naturally hydrophobic surfaces and the hydrophobic moieties of the dextrin molecule plays a significant role in the adsorption process.

b) Presence of charged surface atoms in the talc structure. The weak hydrogen bonds between the charged surface oxygen atoms and the polar hydroxyl groups in the dextrin also contribute significantly to stabilization of dextrin molecules at talc basal plane surfaces.

c) Surface nanoroughness of sulfur at the atomic level. The atomic scale surface roughness produces energy and structural barriers and prevents the simultaneous stabilization of multiple hydrophobic moieties of dextrin molecules at the sulfur surface.

ACKNOWLEDGEMENT The authors are grateful to Professor Jayendran C. Rasaiah for valuable discussions, and grateful for the collaboration with Professor Anh V. Nguyen who helped to initiate the MD simulation research. The financial support provided by the Department of Energy, Basic Science Division Grant No. DE-FG-03-93ER14315 is gratefully acknowledged. In addition, this study was prompted to some extent by collaborative research supported by NSF under grant Nos. INT-0227583 and INT-0352807.

Reference

(1) Fuerstenau, D. W.; Huang, P. *XXII International Mineral Processing Congress* **2003**, 1034.

(2) Balajee, S. R.; Iwasaki, I. *Transactions of the American Institute of Mining, Metallurgical and Petroleum Engineers* **1969**, *244*, 407.

(3) Haung, H.-H.; Calara, J. V.; Bauer, D. L.; Miller, J. D. *Recent Dev. Sep. Sci.* **1978**, *4*, 115.

(4) Healy, T. W. *Journal of Macromolecular Science, Chemistry* **1974**, *8*, 603.

(5) Jenkins, P.; Ralston, J. *Colloids and Surfaces, A: Physicochemical and Engineering Aspects* **1998**, *139*, 27.

(6) Liu, Q.; Laskowski, J. S. *Journal of Colloid and Interface Science* **1989**, *130*, 101.

(7) Liu, Q.; Laskowski, J. S. *International Journal of Mineral Processing* **1989**, *26*, 297.

(8) Liu, Q.; Laskowski, J. S. *International Journal of Mineral Processing* **1989**, *27*, 147.

(9) Miller, J. D.; Laskowski, J. S.; Chang, S. S. *Colloids and Surfaces* **1983**, *8*, 137.

(10) Morris, G. E.; Fornasiero, D.; Ralston, J. *International Journal of Mineral Processing* **2002**, *67*, 211.

(11) Pugh, R. J. *International Journal of Mineral Processing* **1989**, *25*, 131.

(12) Pugh, R. J. *International Journal of Mineral Processing* **1989**, *25*, 101.

(13) Rath, R. K.; Subramanian, S.; Laskowski, J. S. *Processing of Hydrophobic Minerals and Fine Coal, Proceedings of the UBC-McGill Bi-Annual International Symposium on Fundamentals of Mineral Processing, 1st, Vancouver, B. C., Aug. 20-24, 1995* **1995**, 105.

(14) Rath, R. K.; Subramanian, S.; Laskowski, J. S. *Langmuir* **1997**, *13*, 6260.

(15) Wie, J. M.; Fuerstenau, D. W. *International Journal of Mineral Processing* **1974**, *1*, 17.

(16) Wisniewska, S. K. *Ph.D. Thesis, University of Utah* **2005**.

(17) Du, H.; Miller, J. D. *International Journal of Mineral Processing* **2007**.

(18) Nalaskowski, J.; Abdul, B.; Du, H.; Miller, J. D. *Canadian Metallurgical Quarterly* **2007**.

632g Computational & Kinetic Modeling Of C3 Oxidation Over Bifunctional Catalysts

Martin Sanborn[1], Joseph T. Golab[2], Mark Cesa[1], Jeffery S. Logsdon[1], and James Brazdil[2]. (1) INEOS Nitriles, 3030 Warrenville Rd, Suite 650, Lisle, IL 60532, (2) INEOS Technologies, 150 W Warrenville Rd, Naperville, IL 60563

The selective oxidation of propylene via heterogeneous catalysts is one of the most successful commercial processes to produce acrylic acid (CH2CHCOOH). The application of molecular modeling and simulation technologies can help understand how a catalyst can control the reactivity of the system toward desired products from a process in which multiple products are possible.

AUTHOR INDEX

AUTHOR INDEX

AUTHOR INDEX